DICTIONARY
OF
MECHANICAL ENGINEERING

DICTIONARY
OF
MECHANICAL ENGINEERING

S. B. CHOPRA
B.E., M.Tech.

ANMOL PUBLICATIONS PVT. LTD.
NEW DELHI-110002 (INDIA)

ANMOL PUBLICATIONS PVT. LTD.
4374/4B, Ansari Road, Daryaganj
New Delhi-110 002

PRINTED IN INDIA

Published by J. L. Kumar for Anmol Publications Pvt. Ltd., New
Delhi-110 002 and Printed at Mehra Offset Press, Delhi.

Preface

It is a matter of great satisfaction that the first edition of this dictionary got exhaused within a short period. Now I am presenting the new revised and enlarged second edition. This has been revised to make it up-to-date.

This dictionary has been compiled to cover a very large number of mechanical engineering terms in a handy and compact manner. This dictionary also includes terms from various fields which are allied to the mechanical engineering industry such as foundry practice, metallurgy, and welding, etc. The selection of technical terms has been done mainly on the basis of reading of current literature, wherever necessary, line diagrams are also added to make the terms more clear.

The dictionary will be of immense value to the diploma, degree and A.M.I.E. students, and to the mechanical engineers of long experience and having wide interest.

In compiling a dictionary of this kind, it becomes essential to refer the works of many experts in this field and seek the advice of friends and colleagues to all of whom the editor is deeply indebted.

S.B. Chopra

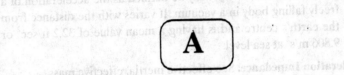

A

a. Refers to the symbol for acceleration. Also f.

A. Refers to the symbol for Ampere.

α. Refers to the symbol for angular acceleration.

ATE. Abbreviation for Automatic Test Equipment.

abs. Abbreviation for absolute.

acc. Abbreviation for acceleration.

Ablation Shield. Protection shields. These are attached to bodies designed to re-enter the earth's atmosphere which vaporize in the intense heat and thereby keep the main structure at a safe temperature.

Abradant. Refers to a material, such as emery and generally in powder form. It is used for grinding.

Abrasion. Refers to the wearing or rubbing away of a surface.

Abrasion Test. Refers to a test by means of a scratch on a smooth surface of the material tested.

Abrasive. Mineral employed in sharpening, grinding or polishing operations.

Absorption Dynamometer. A dynamometer which measures the work done by absorbing or dissipating the power *e.g.*, by the friction of a brake or as in a Froude brake.

Accelerated Fatigue Test. Refers to a fatigue test in which the alternating stress level during the test gets increased above that expected in service so as to reduce the testing time. This technique is mainly used in assessing equipment performance in a vibration environment.

Acceleration. May be defined as the rate of change of velocity (speed) or the average increase of velocity in a unit of time. It is usually expressed in feet (or centimetres) per second per second.

Acceleration Due to Gravity. May be defined as the acceleration of a freely falling body in a vacuum. It varies with the distance from the earth's centre and is having a mean value of 32.2 ft/sec² or 9.806 m/s² at sea level.

Acceleration Impedance. See effective inertia, effective mass.

Accelerator (accelerator pedal)

(a)   It is a pedal in a motor-vehicle which acts on the throttle valve and thereby controls the power and speed of the engine.

(b)   A pedal which is able to control the fuel injection into an oil engine.

Accelerometer. An instrument which is used for measuring acceleration

(a)   By the movement of a mass supported by a spring,

(b)   By a simple pendulum in which its period varies inversely to the square root of the acceleration being measured, or

(c)   By the precession rate of a pendulous gyroscope.

Accessory Gearbox. Refers to a gearbox, driven by and remote from, an engine for mounting accessories like an hydraulic pump on an aero-engine.

Acid Pump. Refers to a pump with the barrel and valves made of glass so as to remain resistant to acids.

Ackerman Steering. Refers to the arrangement on an automobile whereby the inner axle gets moved through a greater angle than the outer axle during cornering so as the provide approximate true rolling of the respective wheels about the turning point. Figure 1 shows that the variation in turning angle of the two

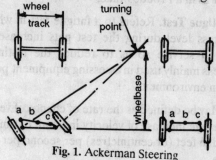

Fig. 1. Ackerman Steering

wheels could achieved by a rod *a b c* which is smaller in length than the distance between the front wheel pivots.

Acme Thread. Screw thread of American origin, the section of which is a mean between the square and vee threads. Used extensively for feed screws. The flanks have an inclined angle of 29°.

Actuator

(*a*)　Refers to an electric, hydraulic, mechanical or pneumatic device, or combinations of these, to effect some predetermined linear or rotating movement. An example used in automatic train control has been an air operated differential piston and mechanism which is mounted on an automatic brake valve to operate at a predetermined brake pipe pressure reduction.

(*b*)　A servomotor which is producing a limited output motion.

(*c*)　A complete self-contained servo-mechanism which is producing a limited output motion.

Adapter. Appliance by means of which objects of different sizes will interchange on a spindle or other fitting.

Adaptive Control. Refers to a type of control in which an original programmed motion gets modified in response to information provided by a sensor system.

Addendum. The radial distance between the pitch circle of a gear wheel and the tip of a tooth. In the British Standard form of tooth it is equal to 0.3183 × circular pitch.

Addendum Angle. Refers to the difference between the tip angle and the pitch angle of a bevel gear.

Addendum Circle. In designing wheel teeth, the circle that passes through the tips of the teeth.

Addendum (screw threads). Refers to the radial distance between the major and pitch cylinders (or cones) of an external screw thread; also, refers to the radial distance between the pitch and minor cylinders (or cones) of an internal thread.

Adhesion (adhesive force). Refers to the frictional grip between two surfaces in contact, *e.g.*, between the locomotive driving wheel and the rail where it has been the product of the weight on the wheel and the friction coefficient (0.1 to 0.2 depending on the condition of the rail surface).

Adhesives. In general, glues which are used for their structural strength.

Adjustable Pitch Propeller. Refers to a propeller, the pitch of whose blades can be changed on the ground but not in flight.

Adjusting Rod. A rod with an adjustable clamp to attach to a fusee or barrel arbor and having with sliding weights for balancing the pull exerted by the main spring and thereby testing its pull.

Adjusting Screw. A screw having a very find thread, in an instrument or tool by which one part is moved relative to another, to provide adjustment in focus, level, tension etc.

Adjustment Strips. Metal strips which are employed for the accurate adjustment of the exact bearing loads on sliding surfaces. The precise amount of contact gets effected by pressure imparted to the strips from set or adjusting screws.

Admission. The term used for the instant in the working cycle of an internal-combustion or stream-engine when the inlet valve permits entry of the working fluid into the cylinder.

Admission Corner. Refers to the corner on an indicator diagram which is corresponding with the entry of the working fluid into a cylinder.

Admission Line. Refers to the side of the indicator diagram which is showing the actual condition while stream is entering an engine cylinder.

Admission Port. Refers to the passage by which stream or the combustible fluid is entering an engine cylinder.

Admittance. See mechanical admittance.

Advance. In an internal combustion engine cylinder, advance occurs when the timing of the ignition is altered to cause the spark in the cylinder to pass at an earlier point. Only done when running at high speed to allow the explosion time to take effect when the piston has reached the top of its stroke.

Advance of Spark. In ignition refers to the turning the contact breaker so that the spark will ignite the charge earlier during the compression stroke.

An internal-combustion engine is having 'advanced ignition' if the spark has been advanced.

Advanced Ignition. See advance of spark.

Advanced Passenger Train Suspension. The arrangement of the Advanced Passenger Train tilt suspension for allevation of side load on bends has been deficted in Figure 2.

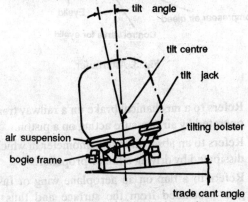

Fig. 2. Advanced passenger train.

Aerial (cable) Railway (aerial ropeway). The term used for a system of overhead cables having small cars or containers which is used for conveying persons or loads, usually over mountainous country.

Aero-engine. Refers to the power unit of an aircraft, which also includes originally piston-engine types and later the gas-turbine types after these had been invented.

Aero thread Insert. See thread insert.

After Burner. Refers to that part of a turbojet engine where extra fuel has been burnt between the gas turbine and the nozzle.

Agitators. Refers to mechanical stirrers or reverberators which are used for settling concrete, sorting coal or sand, or for mixing molten metals or for rocking fluid paper-pulp on a frame to make the wood fibres interlace.

Air Bearing. A shaft bearing which is maintained wholly by compressed air having no contact between fixed and moving surfaces.

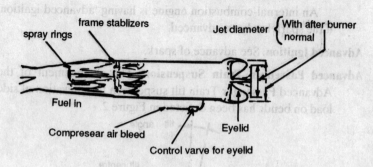

Fig. 3. Afterbufer.

Air Brake

- (*a*) Refers to a mechanical brake on a railway train which gets operated by air pressure acting on a piston.
- (*b*) Refers to an absorption dynamometer in which power gets dissipated by driving a fan or propeller.
- (*c*) Refers to a flap on an aeroplane wing or fuselage which gets protruded from the surface and thus reduces the speed of the aeroplane.

Air Compressor. Refers to a machine which compresses air after it is drawn in at one pressure and delivers it at a higher pressure, as a source of power or for ventilation. It could be of a reciprocating, rotary or fan type.

Air-cushion. A cushion of air which is compressed on the far side of the piston of a high-speed single-acting stream-engine, both to absorb shocks at the release end of the cylinder and to help the engine past the dead centre.

Air-cushion Vehicle. A craft riding on a cushion of air over land or water, the air pressure have been maintained beneath the vehicle by power-driven rotors or fans.

Air Dashpot (See Figure 4 and moving-iron instrument). Refers to an instrument having a loosely fitting piston in a cylinder which permits the slightly compressed air to by-pass the piston and thereby to slow down the motion of the indicating pointer.

Air Drill (windy drill). A rock drill which is operated by compressed air.

Air Ducts. The term used for pipes, drilled holes or cast channels through which air is passing for cooling, heating or supplying a pneumatic device.

Air Ejector. Refers to an air pump which is able to maintain a partial vacuum in a vessel by a high-velocity steam jet to entrain the air and exhaust it against atmospheric pressure.

Air Engine

    (*a*) A heat engine in which air finds use as the working substance. It is not practicable except for very small powers.

    (*b*) A small reciprocating engine which is driven by compressed air.

Air Gap. The space between the armature and the magnets of an electric motor.

Air-gap Torsion Meter. Refers to a measuring instrument in which torsion of a shaft under test makes the air gap between two electromagnets to change thereby altering the current flowing through an ammeter connected to them.

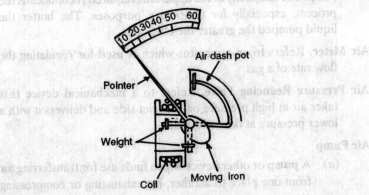

Air Daspot (moving iron instrument).

Air Hammer

    (*a*) Refers to a double-acting power hammer which is used in drop forging for roughing out heavy forgings in foundry work. The power of the hammer has been based only on the weight of the tup and no account has been taken of the additional power provided by the compressed air usually operated at a pressure of from 400 to 500 kPa (60 to 80 lb/in^2). If stream has been used, it has been usually at a pressure from 525 to 700 kPa (75 to 100 lb/in^2).

(b)   Refers to a pneumatic hammer in the form of a pistol. It is used for rivetting.

Air Lift. Machine by which water is lifted from deep wells by the aid of compressed air.

Air-lift Pump.   Pump which acts by forcing air from an air-compressor down a small pipe, which is passed down a well or bore-hole from which water is to be raised. The air pipe is bent upward at its lower end so as to discharge into the rising main. The air bubbles thus formed reduce the density of the mixture as compared with that of the water in the well or bore-hole, and in consequence of the difference in weight between the aerated water in the rising main and that outside it discharge takes place, the amount depending upon the depth to which the main pipe is immersed below the level of the water surface in the boring or sump. The simplicity of the apparatus required recommends the process, especially for temporary purposes. The hotter the liquid pumped the greater the efficiency.

Air Meter. Refers to an apparatus which is used for regulating the flow rate of a gas.

Air Pressure Reducing Valve. Refers to a mechanical device that takes air at high pressure on the inlet side and delivers it with a lower pressure at the outlet side.

Air Pump

(a)   A pump or other device which finds use for transferring air from one place to another, for exhausting or compressing air.

(b)   A reciprocating pump which is fitted to condensing steam-engines to draw water from the condenser together with any vapour or air which is liberated in the process.

(c)   A vacuum pump which is to reduce the pressure on the low-pressure side of a system.

(d)   Refers to a blower to get a rapidly moving air blast.

Air Screw. See propeller.

Air-speed Indicator (ASI). An instrument giving the speed of the aircraft through the air, its reading has to be corrected for instrument error, position error and compressibility error.

Air Standard Cycle. Refers to standard cycle of reference which is used for comparing the performances and calculating the relative efficiences of different inter-combustion engines.

Air Standard Efficiency. Refers to the thermal efficiency of an internal-combustion engine which is working on the appropriate air standard cycle.

Air Starting Valve. A small piston valve in a diesel engine which is actuated by the camshaft and operating the main valve to allow starting air to the working cylinders.

Air Spring (pneumatic spring). Refers to a spring in which compression of air within a cylinder or rubber bellows gives a progressive resisting force.

Air Valve. Valve attached to the summit of a pipe line to discharge automatically any air which accumulates there so as to prevent an air-lock.

Air Vane. See vane.

Air Vessel. A vessel having air which has been fitted :

    (*a*)   To the delivery side of a reciprocating water pump to smooth out the pulsating discharge, or

    (*b*)   To promote an even flow in long pipe-lines.

Airless Injection. See solid injection.

Alighting Gear. Refers to the part of any aircraft which is able to support it on land or water and absorbs the shock of landing, but excluding the hull of a flying boat. It is including all under carriage units of land-planes and the main and wing-tip floats of seaplanes.

Alignment. Means a setting in line of several points like the centres of a lathe the centres of the bearings of an engine crankshaft and the exial continuity of shafting and shaft- bearing.

Alignment Chart. Graphical means of obtaining quickly the results of numerical formula. Also, a chart of showing means of testing, and the tolerances for machine tools.

Allan Valve. See trick valve.

Allen Key. Refers to an L-shaped bar of hexagonal cross-section which is used to tighten up. Allen screws and other types of screw having an internally recessed hexagonal portion to permit tightening.

Allen Screw. A Screw having recessed head.

All-geared Drive. Applied to the growing practice of operating speeds and feeds of machine tools through nests of gears instead of by belts. The gears are enclosed in boxes, which usually form oil wells. The term "all-geared head" is applied to a lathe headstock so fitted.

All-or-nothing Piece (stop side). Refers to a piece of the mechanism of a repeater which either allows striking or entirely prevents it.

Allowance. Prescribed difference between the high limit for a shaft and the low limit for a hole in order to provide a certain class of fit. The allowance may be either positive or negative according to whether a clearance fit or an interference fit, is required.

Alternate Cones. Refers to two equal cones which are arranged on parallel shafts with their bases facing in opposite directions. Their mutual function has been to provide speed variation by means of a shifting belt that can travel from end to end.

Alternating Stress. Refers to the stress in a material induced by a force which is applied alternately in opposite directions.

Ampere (A). May be defined as that constant current which, if maintained in two straight parallel conductors of infinite length, of negligible circular cross-section, and kept 1 m apart in vacuum, would produce between these conductors a force equal to 2×10^{-7} newton per metre of length.

Amplifier. Refers to a device in which an input controls by hydraulic pneumatic or electrical means a local source of power to produce an output greater than, and bearing a definite relationship to, the input.

Torque amplifier (capstan amplifier) A mechanism, having input and output shafts which are rotating at the same speed on the principle of a capstan, to give an amplified output torque when an input torque gets applied; the additional energy is supplied by the rotating capstan drum.

Anemometer. An instrument which is used for measuring and registering the velocity and direction of the wind or the rate of flow of a gas, generally by mechanical or electrical methods.

Aneroid Barometer. An instrument which is usually portable and is used for recording changes in atmospheric pressure and for the determination of altitude. Its construction is based on the principle of a vacuum chamber unit with a train of levers to magnify the amount of the expansion and construction of a bellows type unit. The zero has to be set to the correct sea-level atmospheric pressure at the time so that its reading should indicate the correct altitude.

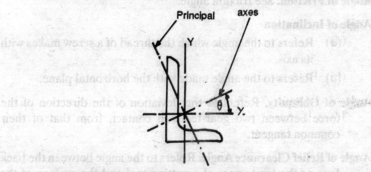

Fig. 5. Principal axes of an angle iron..

Angle (angle bar, angle iron, angle steel). Refers to a structural member, of wrought iron or mild steel bar rolled or extruded which is having a cross-section like the letter L (See Figure 5).

Angle Bearing. Refers to a shaft-bearing in which the joint between the base and the cap is set at an angle, and thus not perpendicular to the direction of the load.

Angle Bending Machines. Used for dealing with other sections besides angles. They embrace two types, one being that in which the bars are bent to various curvatures by means of rolls, the other in which they are squeezed into angular forms in presses.

Angle Bevelling Machine. One is which angles are set between rollers to acute or abtuse sections. The angle may be uniform, or be varied from one end to the other.

Angle Cutter. Milling cutter used for milling flutes on tapes reamers, spiral mills and other cutters.

Angle Gear. Refers to an arrangement of bevel gearing to drive a shaft, at other than a right angle with the driving shaft, by the interposition of a third mitre wheel.

Angle Motion. Canting motion.

Angle of Contact. Refers to the angle which is subtended at the centre of a pulley from the extreme parts of the periphery that are contacted by the belt attached to the pulley.

Angle of Flexure. Refers to the angle through which torsion deflects a shaft.

Angle of Friction. See friction angle.

Angle of Inclination

    (*a*)   Refers to the angle which the thread of a screw makes with its axis.

    (*b*)   Refers to the angle made with the horizontal plane.

Angle of Obliquity. Refers to the deviation of the direction of the force between two gear-teeth in contact, from that of their common tangent.

Angle of Relief Clearance Angle. Refers to the angle between the back face or the lower part of a cutting tool and the surface of the material which is being cut.

Angle of Threat. See included angle.

Angle of Twist. May be defined as the angle through which one section of a shaft gets twisted by a torque, relative to some other section.

Angle of Upset. Refers to the angle at which a portable-type balance crane will upset or overturn with the weight of its load.

Angle Valve. A screw-down stop valve having the casing or body of a spherical shape. The axis of the stem happens to be in line with one body end and at right angles to the other.

Angledozer. See buildozer.

Angles of Cutting Tools (cutting angles). Refers to the angles between the surfaces of the materials being cut and the cutting faces to the tools. The chip thickess gets varied with the approach angle of the tool. (Fig. 6).

Angular Acceleration. Refers to the rate of increase of rotational (angular) velocity. It is expressed in radians per second per second.

Angular Advance (angle of advance)

1. In steam-engine valve gear the angle which the centre of an eccentric sheave is making with a line set a $90°$ in advance of the crank pin. Its magnitude has been found to depend on the lead and the outside lap.

2. Refers to the angle between the position at the time of ignition and the outer dead centre which has been used to optimize the combustion of fuel in spark-ignition engines.

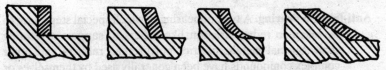

Fig. 6. Chip thickness variation with approach angle.

Angular Cutter. A milling cutter which is having the cutting face at an angle to the axis of the cutter.

Angular Displacement

(*a*) Refers to the angle turned through by a body about a given axis, measured in degress.

(*b*) Refers to the angle turned through by a line joining a fixed point to a moving point.

Angular Momentum. See momentum.

Angular Thread. See vee-thread.

Angular Velocity. May be defined as the rate of change of angular displacement, generally expressed in radians per second or revolutions per minute.

Anisotropic. Not isotropic. A material whose elastic properties vary in different directions same as aeolotropic.

Annealing. Refers to the maintenance of a known temperature for a given time for reducing the number of dislocations within a material.

Annular Gear. An annular ring having gear-teeth cut on it.

Annular Seating. Refers to a ring-shaped seating for a valve. It is as found in pumps.

Annular Valve (circular disc valve). A valve which consists of a circular disc seating on a concentric hole.

Annular Wheel. A cog-wheel in which the teeth are fixed to its internal diameter; also called an internal wheel. It always revolves in the same direction as its pinion.

Annulus. A flat circular part having a concentric circular hole, like a washer.

Anthropomorphic Configuration. See arm and elbow configuration.

Anti-backlash Gear. See gear.

Anti-friction Bearing. A type of bearing in which special steps have to be taken to reduce friction like rollers to support a rotating shaft. Special metals, plastics polyurethan rubbers and other complex compounds have been generally used by themselves or impregnated in the material of the main bearing bush.

Anti-friction Metals. Previously, this term was used to describe white metal, a tin-base alloy containing over 50% tin, and now this term is applied to a wide range of metals which are specially suitable for bearings, especially tin-lead alloys.

Anti Friction Rollers. Live rollers which are able to sustain the pressure of a rotating spindle or shaft.

Antinode. See nodes.

Antinous Release. Refers to a flexible release cable in a camera for operating the shutter.

Antiphase. When the difference in phase angle has been π.

Anti-resonance. When a small change in the frequency of an externally-applied excitation is able to make an increase in the amplitude of a specified response of a mechanical system.

Anti Roll Bar. A steel bar which gets fitted transversely to the body of motor vehicles being curved at each end and attached to the steering wheel suspension. It has to be installed to reduce the roll of the body as the vehicle takes a corner.

Anti-static Belting. Refers to belting with an extremely high coefficient of friction on a highly conductive traction face which is not able to clog fluff etc.

Anti-torque Rotor. See tail rotor.

Anti-vibration. See vibration and mounting.

Anvil (anvil block)

    (*a*) Refers to a massive block of cast or wrought-iron, sometimes steel faced, on which work gets supported during forging.

    (*b*) The jaws of a micrometer are also termed as anvils.

Aperiodic. Having no natural frequency, not resonant at any one frequency.

Apex

    (*a*) Refers to the common intersection of the axes of a pair of bevel gears and the instantaneous axis of relative motion of either gear with respect to the other, termed as the pitch element, which all lie the axial plane.

    (*b*) Refers to the corner of the fundamental triangle opposite to its base in the geometry of screw thread.

Approach Angle. See angles of cutting tools.

Apron. A plate or fixing which is bolted to the front of the saddle of a lathe. It encloses the gear operated by the lead screw.

Apron Conveyor. A travelling belt which is composed of a number of linked sections. It is usually metal or wood slats, for transport horizontally or on a gentle gradient; also called 'salt conveyor'. It has been used on large machines to protect the bed from swarf. Aircraft wing spar milling machines often incorporate this type of apron.

Arbor

    (*a*) Refers to a rotating shaft, spindle or bar which forms part of an instrument or machine of machine tool.

    (*b*) Refers to a spindle of a wheel as in a watch or clock.

    Expanding arbor. A lathe arbor than can get expanded by blades sliding longitudinally in taper keyways and used for supporting work-pieces of different bores.

Arbor Chuck. A chuck which is used in a lathe for turning the outside diameters of cylindrical work after the hole is first bored, the hole fitting over an arbor, mandrel or spindle by ensuring concentricity of outer and inter diameters.

Arbor Press. An appliance which used for forcing arbors or mandrels into or out of work by a screw-press or hydraulic power.

Arboring. Signifies the shouldering back of a flat bearing face, to receive the washers and nuts of attachment bolts. It is done by means of a broad facing cutter wedged transversely in a boring bar, or arbor. Where fillets or radii intersect flange faces, and where the surface of a casting is from any cause uneven, arboring is usually and properly resorted to.

Archimedean Drill (persian drill). A drill having a quick multiple thread over which a nut works in a to-and-fro axial movement to provide an alternating rotary motion of a bit.

Archimedean Screw. Refers to a hollow inclined screw, or a pipe which is forming a helix around an inclined axis, with its lower end in water. When it gets rotated the water gets lifted to a higher level. (Figure 7).

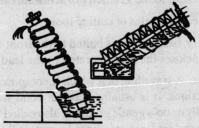

Fig. 7. Archimedean screws for raising water.

Archimedean Spiral

(a) Refers to the locus of a point which is moving with uniform velocity along the radius vector while the radius vector also moves about the pole with constant angular velocity. In polar co-ordinates $r = a\theta$.

(b) Refers to a device for raising water by rotation of a spiral.

Arc Welding. Method of welding in which the metal is fused by the heat of an electric arc. The process can be considered under four headings : metallic arc welding; carbon-arc welding; argon or helium shielded arc welding; and atomic hydrogen welding. Each process is described under its respective heading.

Armature. Refers to the shaft and rotating attachments of a direct current electric generator or electric motor.

Articulated Blade. Refers to a rotorcraft blade which is mounted on one or more hinges to allow flapping and fore-and-oft movement during flight.

Articulated Connecting-rods. Refers to the auxiliary connecting-rods of a radial engine which are working on pins carried by the master-rod instead of on the main crank pin.

Artificial Horizon (Gyro Horizon). Refers to an instrument embodying a gyroscope which is able to stimulate the natural horizon.

Artificial Intelligence. Refers to the ability of a machine to carry out certain functions associated with human intelligence, *e.g.,* interpreting data, solving problems and decision making.

Assembly. Refers to the putting together of a machine, or mechanism, from its component parts; also the final product after putting the parts together.

Subassembly. Refers to any part or parts of an assembly which can be treated as a separate item.

Atkinson Cycle. Refers to an internal-combustion engine cycle in which the expansion ratio exceeds the compression ratio.

Atmospheric Engine

    (*a*) Refers to an engine of the piston type which is not supercharged.

    (*b*) Refers to an early form of steam-engine, in which a partial vacuum created by condensation of the steam allowed atmospheric pressure to drive down the piston; also termed as a single-acting engine.

Atmospheric Line. Refers to a datum line on an indicator diagram which is drawn by allowing atmospheric pressure to act on the indicator piston or disphragm; it divides the steam area above from the vacuum area below.

Atmospheric Pump. Suction pump.

Atwood Machine. A device which consists of a pulley over which is passed an inextensible cord connecting two weights. It can be used for determining acceleration of gravity.

Auger. A tool which is used for boring holes. It consists of a long steel shank with a cutting edge at one end and a cross-piece for handle at the other. A 'shell' or 'pud-auger' possesses a straight channel groove. A 'screw auger' has a twisted blade, the chips being discharged by the spiral groove.

Auger Stem. Refers to the heavy bar to which the drill bit gets attached when boring a well.

Autoclave. Refers to an oven, in which a pressure can be produced. It is used to heat plastic and reinforced plastic items and bonding materials to ensure that they are fully cured.

Auto-collimator. See optical tooling.

Autographic Diagram. Diagram made on squared paper by the machine to which it is attached and used in connection with engine performances, steam indicators, in testing machines, furnaces electrical apparatus, etc.

Automatic Expansion. Refers to the control by governors and their gear of the expansion of steam in a steam-engine.

Automatic Lathe. A lathe which is having an unmanned repetitive action.

Automatic Pilot. See autopilot.

Automatic Stoker. See mechanical stoker.

Automation. A technique which is used for controlling the whole or a part of a manufacturing process, including inspection and rejection. Part or all of the technique has been automatically under electronic control.

Automobile. A mechanically propelled vehicle running on common roads. Steam, electricity, alcohol, and petroleum spirit are the agencies employed, the last-named vastly predominating. The manufacture of the engines and gears for these vehicles has occasioned the design of a large number of new and modified machine tools, and has greatly influenced shop methods.

Autopilot. The term used for the mechanism and its associated controls for controlling automatically the flight of an aircraft or a missle along a given path.

Auxiliary Rotor. Refers to a small rotor which is mounted on the tail of a helicopter of provide directional control and tocounteract the torque of the main rotor, generally called tail rotor.

Axial (axial-flow) Compressor. Refers to a compressor having alternate rows of fixed and rotating blades, radially mounted, with the flow through the compressor in the direction of the axis. Two main types are known, *i.e.*,

(*a*)   with tapered casing, and

(*b*)   with tapered rotary drum.

Axial Engine

(*a*)   A turbojet engine having an axial-flow compressor.

(*b*)   A piston engine having cylinders paralled with the driving shaft.

Axial Pitch. The pitch of a screw of gear which is measured in a direction parallel to the axis.

Axial Plane. A plane having the axis of a symmetrical body.

Axial Pressure Angle. See pressure angle (axial).

Axial Section. Refers to a section in a plane having the axis of a screw or gear.

Axial Thickness. Refers to the distance measured along a line parallel to axis between the traces of a gear-tooth that is, the distance which is measured across the reference cylinder of a helical, spur of worm gear-wheel in a direction parallel to the axis.

Axial Turbine. Refers to a turbine in which the water or gas is passing through the wheel in the axial direction.

Axis (or rotation). Refers to a straight line about which a body, a screw or gear rotates.

Axle. Refers to a cross-shaft which is carrying the driving or freely-mounted wheels of any vehicle. In the 'dead axle' type the wheel turns on the axle which gets inserted in the hub and forms the aixs of rotation. In the 'live axle' type the wheel is rigidly fixed to the axle that turns in bearings.

Axle-box. Refers to the complete bearing arrangements for the axles of railway rolling-stock with the upper half of the bearing having a box-shaped housing to hold the lubricant.

Axle-grinding Machine. A machine which is used for grinding railway axles with the wheel on the axles.

Axle Lathe. A lathe usually double, that is, having a loose poppet at each end of the bed, between which the axle is centred. It is rotated through a central headstock with double drivers, through gears with speed changes. Two slide-rests deal with the axle journals.

Azimuth Control. Refers to cyclic pitch control.

B

BHN. Brinell Hardness Number.

BHp. Brake Horse-power.

BM. Bench Mark; Bending Moment.

BMEP. Brake Mean Effective Pressure.

Babbiting. The term used for the process of lining bearings with Babbitt's metal or with white metal.

Babbitt's Metal. An alloy which is used for bearings containing tin alloyed with copper and antimony plus varying amounts of lead. A common formula has been tin 40, copper 1.5, antimony 10 and lead 48.5 as percentages.

Babcock and Wilcox Boiler. Refers to a watertube boiler which is basically a horizontal drum from which a pair of headers gets suspended to support the ends of a bank of straight tubes.

Babcock and Wilcox Mill. A dry grinding mill which is using rotary steel balls.

Back Centre. Refers to a pointed spindle on the loose headstock of a lathe which is used for supporting the end of the work remote from the chuck.

Back Cock. Refers to the bracket on the back plate of a clock from which the pendulum gets suspended.

Back Cone. Refers to the cone whose generator has been perpendicular to the pitch cone generator at the pitch circle of a level gear.

Back Cone Angle. Refers to the angle between the axis and the pitch angle.

Back Cone Pressure Angle. Refers to the acute angle between the normal to the intersection of the tooth flank of a bevel gear and the back cone at the pitch circle and the tangent to the pitch circle at the point.

Back Cut-off Valve. Refers to a sliding and adjustable plate on the back face of the main slide valve of a stream-engine which is used to regulate the point of cut-off for the steam and worked independently from a separate eccentric.

Back-firing

(a) The term used for a premature ignition in an internal-combustion engine before this end of the compression stroke, with the consequent reversal of the direction of rotation during starting.

(b) Refers to the ignition of gases while the exhaust valve has been still open.

Back Gear. Refers to a train of gearwheels which are fitted to the headstock of a lathe or other machine tool for the reduction of the speed of the mandrel below that of the cone pully, thereby increasing the power of the machine.

Back-kick. The term used for describing the violent jolt backwards which is felt when an internal combustion engine fails to start due to back-firing.

Backlash

(a) Refers to the amount an element of a mechanism has to move before communicating its motion to a second element.

(b) For two gearwheels, backlash refers to the minimum distance between the tooth flanks which are in mesh.

Back Pressure

(a) Refers to the pressure opposing the motion of a piston during the exhaust stroke, or working stroke, in an internal combustion engine or steam-engine.

(b) Refers to the exhaust pressure of a turbine.

Back-pressure Engine. Refers to a steam-engine in which the steam gets exhausted for heating purposes at a pressure greater than the normal terminal pressure.

Back-pressure Turbine. Refers to a steam-turbine from which the whole of the exhaust steam has been taken at a suitable pressure for heating purposes.

Back-pressure Valve. A valve which is used to prevent the return flow of fluids in a pipe.

Back-rest

(a)   Refers to a guide attached to the slide rest of a lathe and placed in contact with the work to steady it when turning.

(b)   Refers to the roller or oscillating bar at the back of a loom over which the warp threads pass from beam to healds.

Back Shaft. Refers to the shaft which runs along the whole length of the rear of a self-acting lathe and through which motion is transmitted from the headstock to the slide rest for sliding and surfacing only; it has been capable to reversal for traversing the saddle.

Backward Eccentric. The eccentric which is able to open the slide valve to the steam supply when the engine is needed to run backward. Sometimes termed as backward gear.

Backward Gear. Refers to the relative arrangement of eccentrics, etc., in a steam-engine whereby the engine will, on the admission of steam, run backward.

Back Washer. A machine which is used for scouring, drying and opening out carded silvers in worsted manufacture.

Backing-off

(a)   Refers to the operation of relieving or bevelling off the backs of the teeth of milling cutters.

(b)   Refers to the operation of bevelling the hinder or leaving edge of the threads of a tap.

(c)   Refers to the reversal of the spindles of the mule in cotton spinning, to unwind the yarn after the completion of twisting and drawing out.

Backing-off Lathe. Refers to a lathe in which the teeth of milling cutters and taps have been bevelled off by a to-and-fro movement of a cutting tool on the slide rest.

Baffle Plate. A plate which is used to direct the flow of a fluid into a desired direction.

Balance Arc. Refers to the portion of the vibration of the balance of a watch or clock during which it is an contact with the escapement.

Balance Arm. Refers to the portion of the balance which is connecting the rim to the staff.

Balance Box. It is a box for a cantilever-type crane having a heavy load to counterbalance the weight of the jib and the crane's load.

Balance Cock. The detachable bracket which is carrying the upper plot of the balance staff.

Balance Cylinder (balancing cylinder). Refers to a small auxiliary steam-cylinder which is sometimes fitted to large vertical steam-engines to reduce the load on the valve gear by admitting steam to the underside of the balance (dummy) piston which gets connected to the engine slide valve.

Balance Gear (US). Refers to the differential gear of motor- vehicle.

Balance Turning Tool. A very rugged tool which is used for making roughing cuts. It comprises two tool with bits designed to cut tangentially on opposite sides of the workpiece that is being turned.

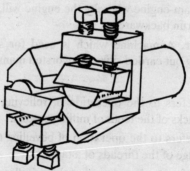

Balance turning tool

Balance Wheel
  (*a*)   A flywheel.
  (*b*)   A spring-controlled and dynamically-balanced wheel which is able to regulate the beats of a watch or chronometer by its oscillations; a balance.

Balanced Druaght. Refers to the system of air-supply and extraction to a boiler or furnace where the grate pressure remains at atmospheric pressure.

Balanced Valve. An equilibrium valve.

Balanced Wheel

(a) A rapidly rotating wheel having a truly-turned rim and holes drilled in or near the rim so that the wheel turns freely and comes to rest in any position.

(b) A wheel which is having dynamic balance.

Balancing Machine. A machine which is used for testing static balancing and determining the weight and position of the masses to be added to obtain balance.

Baler. A machine which is used for compressing loose bulky material and securing it in a convenient form of transport, such as hay or cotton.

Ball-and-socket Joint. A joint in which a spherical end is kept within a socket that has been recessed to fit it, thus allowing free motion within a given cone or cut-out in the socket. It is the same as 'ball joint'.

Ball Bearing. Refers to a bearing on a shaft composed of a number or hardened-steel balls which are rolling between an inner race forced on the shaft and an outer race carried in a housing. The balls are equally spaced by a light metal cage and run in shallow grooves called ball-tracks. The cage or ball retainer gets crimped or riveted into place after the balls have been inserted (see Fig. 1). A self-aligning ball-bearing is having the balls running in a spherical housing which enables the inner and outer races to be at an angle to each other as shown in Fig. 2.

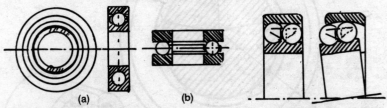

(a)   (b)

Fig. 2. B-2. Ball-bearing, single row; (a) radial or journal, (b) thrust.

Fig. 2. B-3. Ball-bearing, double-row self-aligning-radial.

Ball Bush. An outer cylindrical sleeve which is running on balls along a shaft, each row of balls taking the load in turn and then recirculating as shown in Figure 3.

Ball Catch. Refers to a spring-controlled ball, projecting through a smaller hole, which engages with a hole in a striking plate, as for a door fastening.

Ballcock. A self-regulating valve which, through a linkage system, turns the flow of water (or a liquid) on and off by the falling and the rising of a partly submerged sphere, usually a hollow ball.

Ball Cutter. A spherical cutting tool or a cutter having a rounded edge.

Ball Mill. Refers to a fine grinder or ore crusher which has been a slightly inclined or horizontal rotating cylinder containing balls, usually ceramic, or steel, to grind the material to the necessary fineness by the rubbing and impact of the tumbling balls. Wet ball-milling has been generally a batch process, but dry ball-milling may be continuous with the 'fines' removed by an air curent.

Ball Nut. Refers to a nut having a semi-circular helical groove on the inside which fits over a shaft with a mating groove. The load gets transmitted by balls running in the grooves and returning through a non-load-carrying section. (See Fig. 3).

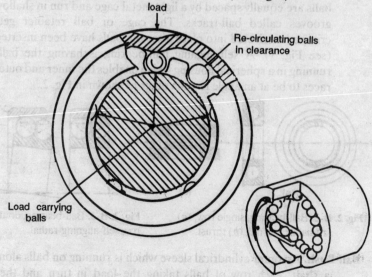

Fig. 3. (Above) Ball bush, Fright Ball nut.

Ball Race

    (*a*)   A steel ring which is forming part of the ball-track of a ball-bearing.

    (*b*)   A complete ball-bearing.

Ball (spherical) Resolver. A sphere having a fixed centre, two degrees of rotational freedom and two output rollers in the equatorial plane with their axes at right angles to each other. The driving-roller axis has been in a plane parallel to the equatiorial plane and its projection on the equatorial plane has been making an angle θ with the cosine output roller. For a rotation α of the driving-roller output rollers have been rotating through angles equal to $\alpha \cos \theta$ and $\alpha \sin \theta$.

Ball Turning. Refers to the production of spherical objects by means of a special rest moved by worm gears in a circular path or by means of special curved tools, the tool post swinging through a circular arc.

Ball Valve Cage Valve (*a*), Cock Valve (*b*). A non-return valve consisting of a globe or ball working on a cup-shaped seat which is usually within a suitable cage. The spherical ball with a cylindrical hole through its centre allows fluid to flow. When turned through 90° the face of the ball stops the flow. It has been used in small water-and-air-pumps and for small check valves.

Band Brake. A flexible band, having one end anchored and a force applied to the other, which gets wrapped partially round the periphery of a wheel or drum and pulled tight to slow the wheel.

Band Conveyor (belt conveyor). A travelling endless belt which is used for conveying materials, small articles, etc., from one place to another and passing over, and being driven by, horizontal drums.

Band Mill. Refers to a wide bandsaw.

Bandsaw (bandsawing machine). Refers to an endless band of steel having saw-teeth upon one edge which passes over and gets driven by two pulleys with horizontal axes, the pulleys keeping the band in tension. In one type the band runs vertically, and in another type the machine cuts horizontally.

Banjo Axle. Rear-axle casings is in automobile having a differential casing in the centre, thereby resembling a banjo with two necks.

Banjo Bolt. A bolt which is having a blind hole drilled up the centre and several radial holes drilled into it from a peripheral groove. The bolt has been used as an easy method of attaching a pipe flush with a machined surface. It has been frequently used for gas, lubricant feed and bleed pipes.

Banking Pins. Vertical pins, limiting the motion of the lever, which gets located in the botton plate of a watch.

Bar Gauge. Refers to a substitute for a plug gauge which is used for checking the dimensions of large-diameter plain holes.

Bar Lathe. A small lathe, the bed of which has been made in a single piece of circular, triangular or rectangular section.

Bar Movement. A watch movement having the upper pivots carried in bars.

Bar Saw. A large and robust mechanical hacksaw which is used for cutting iron or steel in bar form; it has a very slow action.

Barker's Mill. A mill which is driven by an excess of hydrostatic pressure, in which water gets admitted to a central vertical container and passes out under pressure at the bottom through side holes in hollow radiating horizontal arms, thus driving the mill.

Barostat

(*a*) A device, similar to an aneroid barometer. It is used for regulating the pressure supply to, or delivery from, the fuel matter of an aero-engine.

(*b*) A device which is used for initiating movement or controlling a mechanism when a given pressure is reached.

Barrel

(*a*) Refers to a cylindrical part of a machine such as the body of a pump in which the piston moves.

(*b*) The cylindrical portion of a locomotive boiler or a portable engine boiler.

(*c*) The central part of a propeller hub which takes the centrifugal force on the blades.

Barrel Cap. Refers to the detachable cover of a barrel in a clock or watch.

Barrel Cover. Refers to the cover which shaps into a grooved recess at one end of the barrel in a clock or watch.

Barrel Elevator. Refers to an elevator using parallel travelling chains having projecting curved arms. The chains pass over sprocket wheels at the top and the bottom and lift barrels in the curved arms a loading platform to a runway.

Barrel Hook. The means by which the mainspring gets attached to the barrel.

Barrel Hopper. A machine which is used for disentangling, orientating, the feeding small items during a manufacturing process. The barrel is revolving tumbling the items on to a slopping, vibrating feeding table.

Barrel-type Crankcase. A petrol-engine cranckcase from which the crankshaft has to be removed from one end.

Barrel Wheel. The last, a large, wheel in the train of gearing of a crane, keyed upon the same shaft as the lifting barrel.

Bascule Bridge. Refers to a counterpoise bridge which can get rotated in a vertical plane about axes at one or both ends, the roadway rising and the counterpoise descending into a pit. Also called 'balance bridge'.

Base Cylinder (of a gear). Refers to the cylinder to which the generators of an involute helicoid are tangent.

Base Diameter. Refers to the diameter of the base circle of a helical spur or worm gear.

Base Pitch (normal). Refers to the distance between similar flanks of two adjacent teeth of a gear measured along a common normal.

Base Pitch (transverse). Refers to the distance between similar profiles of two adjacent teeth of an involute helicoid gear which is measured in a transverse plane along a common normal.

Basic Angle. Refers to the angle size on which the design size of a gear, thread etc., has been based.

Basic Form of Screw Thread. Refers to that form on which the design forms for both the external and internal threads have been based.

Basic Member. A mating part whose design is equal to the basic size.

Basic Size. Refers to the size of dimension, or part, on which both the limits of size and the design sizes have been based.

Basic Truncation (major or minor). The distance which is measured perpendicular to the axis, between the appropriate cylinder, or cone and the adjacent apex of the fundamental triangle.

Basil. Refers to the bevelled edge of a drill or chisel.

Bastard Thread. A screw thread which has been not a standard.

Bath Lubrication. Refers to the lubrication by dipping the part, such as a moving gearwheel or chain, into a bath of oil.

Batten. Refers to the swinging frame of a loom which controls the reed, carries the race board and beats up each pick of weft to the fabric already formed.

Baulk Ring (bloker ring, inertia lock). Refers to that part of a synchromesh gear-changing unit which has been designed to reduce the noise or crashing of the gears due to a quick change. Cone pressure has been proportional to speed of change and here an interception device does not allow positive engagement until the speed of the two members is equal.

Bayonet Engine. Refers to a horizontal engine with the bed plate curved round to one side of the crank, the curved portion carrying the bearing for the crankshaft.

Bayonet Fitting. A socket having spring-loaded base and two diamatrically opposed L-shaped slots at its mouth. A mating plug having a pair of pins on its diameter can get pushed in against the spring and retained by the pins. Many light bulbs have been held in their sockets by this means.

Beading Machine. A sheet-metal tool which is used to make flanges, beads (rounded edges) and miscellaneous odd curves and angles.

Beam

(*a*)   Refers to a grider supported at its ends and loaded transversely.

(*b*)   Refers to the beam on which the sheet of threads has been wound in beam-warping.

Beam Engine. An obsolete type of steam engine in which the connection between the piston of the inverted steam-cylinder and the flywheel or pump-cylinder was done through a beam whose point of oscillation was set midway between the centres of the two rods.

Bearing Lubricants. These lubricants are usually petroleum oils or greases made from these oils; vegetable oils; air or inert gases.

Bearing Materials. Bronzes, white metal, Babbitt's metal, copper, nylon and similar materials, polyurethane, metals impregnated with PTFE (polytetrafluoroethylene).

Bearing Neck. Refers to the portion of a rotating shaft in contact with a bearing.

Bearing Plate. Refers to the part of a lathe bed over which the cross-slide slides.

Bearing Pressure. Refers to the pressure of a rotating shaft on its bearing, usually measured in kPa (or 1 b/in^2) of projected area.

Bearing Spring. The type of spring which carries the weight of a vehicle and lessens the effect of jars and shocks.

Bearing Surface

    (*a*)   Refers to the area of the surface upon which a shaft rotates.

    (*b*)   In machinery, the surfaces of bearing parts in mutual contact.

Bearings. The supports for holding a revolving shaft in its correct position.

Beat

    (*a*)   Refers to a measured sequence of strokes or sounds.

    (*b*)   Refers to a sound of regularly varying intensity due to the combination of two sounds of slightly different frequencies.

    (*c*)   Also, refers to the blow given by a tooth of an escape wheel as it strikes the pallet in the clock. When the blow has been uniform on both pallets, the escapement is said be 'in beat'.

Beat Pins. The pins which are projecting from the ends of the gravity arms of a gravity escapement.

Beat Screws. Screws which are used to adjust the relative position of the crutch and pendulum so that the escapement can be brought into beat.

Beater

(a) A revolving shaft having blades which break up and loosen matted lumps of cotton during the cotton-spinning processes of opening and beating.

(b) A trough having a cylinder, both, fitted with knives, for reducing paper pulp to the right consistency.

Beating. The regular thudding sound of a locomotive or a steam vessel.

Beating-up. Refers to the movements of the reed to push each thread of weft against the edge of the woven fabric.

Bedplate (baseplate). A heavy cast iron or fabricated steel base which is used as a foundation for an engine or other machine.

Bedding (bedding in)

(a) Refers to the adjustment and fitting of the journals of a shaft and its bearings to each other.

(b) Refer to a seating or a bed.

(c) Refers to the laying of a piece of machinery on its foundation.

Beetle. A machine in which a row of wooden hammers falls on a roll of cloth as it revolves and imparts a soft glossy finish to it.

Beetle-head. The monkey of a pile-driver.

Bell-crank Lever. A lever having two arms at right angles and a common fulcrum at their junction.

Bell Hopper. The type of the hopper which releases and spreads out by its bell shape the charge of iron ore, coke and limestone into a blast-furnace.

Bellows

(a) A portable or fixed contrivance which is used for producing a jet of air from a flexible-ended or-sided box which is alternately expanded and contracted, drawing in air through a non-return valve and expelling it through a nozzle.

(b) Also refers to the flexible folding light-tight part of some cameras, uniting the back and front portions.

(*c*)   Also, refers to the convoluted portion of a pipe to allow for thermal elongation or misalignment.

Belt (belting; driving band). Refers to an endless band of leather or other flexible material which is used for transmitting power from one shaft to another by running over flat, convex or grooved rim pulleys. Belts may be flat, vee-shaped or ribbed to fit on to appropriately shaped pulleys. Belt velocities may be as high as 800 m/s (15000 ft^1/min).

Belt Compressor. An air compressor which is driven by a belt and pulleys from an independent engine or from shafting.

Belt Coupling. Refers to the union of the ends of a belt.

Belt Drive. Refers to a method of power transmission from one shaft to another by means of an endless belt passing around a pulley on each shaft.

Belt Fastener. A connecting piece which is able to join together the ends of a belt.

Belt Polisher. A polishing machine which consists of a belt covered with abraisive or polishing material and passing around pulleys.

Belt Punch. A cutting tool shaped like piles having an annular edge which is used for cutting holes in leather or similar material.

Belt Sander. A machine in which a belt covered with abrasive material gets moved rapidly by rotating pulleys over woodwork to finish its surface.

Belt Shifter (belt forok). A forked device which is used for shifting a belt from one pulley to an adjacent pulley or from a fast to a loose pulley and vice versa. Also termed as 'belt striker'.

Belt Tightener

(*a*)   A contrivance which is used for pulling the ends of belts together for coupling up.

(*b*)   A device which is used to maintain a uniform tension upon driving belts or to cause them to conform more nearly to the circumference of the pulleys.

Belt Tripper. A contrivance for tilting sideways, at a convenient point, either a belt or apron conveyor.

Bench Drilling Machine. Refers to a small drilling machine which can get bolted to a work bench and actuated by hand or by an electric motor via a suitable speed reduction mechanism.

Bench Lathe. A small lathe which is mounted on a work bench.

Bench Marks. Fixed points of reference which are used for measuring strains.

Bench Work. Work carried on at a bench, or vice, with hand tools or small machines in contrast to work carried on with a lathe or other machine with its own stand and pedestal.

Bending. Means the curvature of a beam about its axis under load.

Bending Moment. Defined as the sum of all moments of a force acting at a point in a body.

Bending Moment Diagram. Refers to a graphical representation of the moment of a force for each force acting on a body, which shows the bending moments in newton metres for the loads indicated in the loading diagram.

Bending Rolls. Heavy rollers of cast-iron or steel set in strong standards to straighten crooked plates, to bend them into arcs, or to bend them into complete cylinders.

Bentail Carrier. A lathe carrier having a bent shank projecting into, and engaged by, a slot in the driving plate or chuck.

Beranger Balance. Refers to a balance in which the motion of each scale pan has been such that the pan remains horizontal for any vertical displacement. The position of the load on the pan does not influence the equilibrium of the balance and the subsidiary beams and links ensure that there have been no lateral thrusts on the knife edges. The total equivalent downward force at A has been equal to the weight on the pan if CE/CE = FA/FB as in Figure.

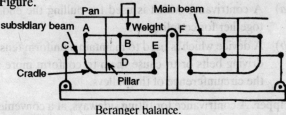

Beranger balance.

Between Centres. In lathe work, this term is used to signify chucking the work between the centres of the headstock and the tailstock. The workpiece then gets rotated by means of a lathe carrier. Other methods involve the use of a chuck, collet or faceplate.

Bevel. A angle, which has been not a right angle, between two surfaces.

Bevel Gear (bevel gearing). An arrangement of bevel wheels for the transmission of motion from one shaft to another on interesting axes.

Bevel Gear Shaper (planer). A machine tool which is used for shaping bevel wheels.

Bevel Mortise Wheel. One of a pair of bevel wheels which is fitted with inserted wooden teeth to secure a silent drive.

Bevel Wheels. Toothed wheels shaped like the frustrum of a cone, which have been used in pairs to transmit motion between two shaft whose axes intersect at an angle at each other.

Beveloid Gearing. An involute gear having tapered tooth thickness, root and outside diameter.

Bezel. Refers to the sloped cutting-edge of a chisel or other cutting tool.

Bib-valve. Refers to a draw-off tap closed by screwing down a washered disc on to a seating.

Bicycle
- (*a*) A vehicle which is having only two wheels, one behind the other.
- (*b*) An aeroplane's undercarriage unit which is having two wheels or wheel units, one behind the other.

Bifilar Suspension. Refers to a suspension by two parallel vertical wires (or threads) to provide a controlling torque on a body, as in some instruments.

Bight. Refers to hanging loop of a chain' or rope which has been held by its ends, such as the operating chain manually-worked pulleys in lifting tackle.

Billet Mill. A rolling mill which is used for reducing steel ingots to billets.

Binary Vapour Engine. Refers to a heat engine which uses two separate working fluids for the high and low-temperature portions of the cycle respectively.

Binder (harvester). Refers to a harvesting machine which, in addition to cutting corn, gathers it, forms in into sheaves and ties the sheaves.

Binder Pulley. It is an adjustable pulley which tightness a belt or cord on its driving and driven pulleys to its correct tension.

Binding Head Screw. A screw head which has been undercut round the steam so that when it has been screwed down tight the peripheral material beds into the mating part.

Binding Screw. A set screw which has been used for clamping two parts together.

Bit

  (*a*) A boring tool which has been rotated in a vice or machine.

  (*b*) Refers to the cutting edge of a borer used in rock drilling. In the USA, the entire length of the borer, including shank, steel and bit.

Bit Stop. An attachment to a bit which limits drilling or boring to a given depth. Also called 'bit gauge'.

Black Bar. The finished product when steel has been produced by hot rolling in a rolling mill.

Black Work. Work that is not machined or polished.

Blacking Mill. Refers to a large revolving closed cylinder having heavy rollers rotating freely on its internal diameter or spherical balls for grinding graphite and carbonaceous materials for the preparation of blacking for painting the inside of casting moulds.

Blade

  (*a*) Refers to the cutting part of some edge tools.

  (*b*) Refers to a vane of a turbine or fan.

  (*c*) Refers to a radial arm of a propeller.

  (*d*) Refers to the movable part of a knife-switch.

  (*e*) Refers to the secondary arm of a square.

Blade Angle. The term used the acute angle between the chord of a section of a propeller, or of a rotor blade, and a place perpendicular to the axis of rotation.

Blank. A specially prepared piece of metal which is ready for machining; grinding; pressing; drawing or extruding, to a particular shape.

Blanking. Refers to a shearing process in which sheet-metal has been cut or punched to form a flat blank, which is later bent on formed in a desired shape.

Blast. Air under pressure which gets impelled by mechanical means.

Blast Engine. An engine which is used for creating a blast of air, generally to aid combustion in a furnace; now largely driven by the waste gases (after cleaning) from the furnace.

Blast Pipe. Refers to the exhaust pipe of a locomotive which is terminating in a blast nozzle. The steam through the nozzle gives a draught to entrain and exhaust the flue gases.

Bleed

    (*a*) Air under pressure which is taken from an axial compressor to run an auxillary service.

    (*b*) Liquid which is taken from a hydraulic system to release trapped air.

Blind Hole. A hole drilled part of the way into a piece of material to any required depth.

Block. Refers to the housing holding the sheaves or pulleys over which a rope or chain passes, as in a lifting tackle.

Block Brake. Refers to a vehicle brake in which a block of metal or hardened material has been forced against the rim of a revolving wheel by hand power or a mechanism.

Block Carriage. The travelling frame which is carrying the chain sheaves upon the horizontal jib for a crane, which gets traversed along the jib by racking gear.

Block Gauge. A distance gauge which is made of hardened steel with its opposite faces parallel and accurately ground flat, the faces being separated by a definite distance – the 'gauge distance'. A block gauge has been used for checking the accuracy of other gauges.

Blocking-up. Refers to the raising and supporting machinery or other construction with the aid of cranes, jacks, levers and wooden blocking.

Blocking Girders. Girders which are attached to the underside of the trunk frames of a travelling crane; fitted back and front, they have been wider than the frames. Jacks fitted at the girder ends are screwed down to the ground to prevent overturning when lifting a load crossways.

Bloom. It is a product in the rolling of steel having a cross-section greater than 36 in^2. Smaller sizes are known as billets.

Blooming Down. Means the rolling down of steel ingots into blooms.

Blooming Mill. A rolling mill which :. .sed for reducing steel ingots to blooms; also known as cogging mills.

Blow Back. Means the return of some of the induced mixture through the carburetter of a piston engine when running at slow speeds due to the late closing of the inlet valve during compression.

Blow-by. Refers to the gas which leaks past the piston of a piston engine during the period of maximum pressure.

Blower

   (*a*)   A rotary compressor which is used for supplying a large volume of air at low pressure.

   (*b*)   A ventilating fan or venturi tube which is used for supplying air.

   (*c*)   A ring-shaped perforated pipe which is encircling the top of the blastpipe of a locomotive.

Blower and Spreader. A machine which is used for combining the action of beaters and blower for spreading cotton into a lap.

Blowing Cylinders. Double-acting cylinders used for pumping the air under pressure into the blast-main of a blast-furnace.

Blowing Engine. Refers to a combined steam-or-gas engine which gets coupled to a large reciprocating air-blower for supplying air to a blast-furnace.

Blowing Through. Means the sending of a jet of steam through the cylinders and valves to warm a steam-engine before starting.

Blueing

(a) Refers to the thermal treatment of watch springs to get the desired elastic properties as indicated by the colour.

(b) Also, refers to the forming a protective coating of a blue oxide film on polished steel by heating in contact with saltpetre or wood ash or incidental to annealing.

(c) The term used for applying a blue dye to metal objects before scratching the dimensions on to the work.

Blunger (blunging machine). Refers to a pottery machine in the form of a vertical cylinder having a rotating shaft armed with fixed horizontal knives. It is used for amalgamating clay with water in making slip, which is clay reduced to the consistency of cream.

Blunt Start. Refers to the condition which gets resulted from the removal of the partial thread at the end of a screwed member to facilitate the entry of the threads without damage, when repeatedly assembled.

Bob

(a) Refers to the suspended weight at the end of a plumbline.

(b) A small buff wheel which is perforted and nearly spherical. It is mounted on a spindle and used for polishing the inside of spoons.

(c) A working beam of a steam-engine.

Bobbin

(a) Refers to a spool on which yarn is wound; the spool is having flanges for holding warp yarn and no flanges for holding the weft.

(b) A small spool which gets adapted to receive thread and installed within the shuttle of a sewing machine so that the thread produces the lower half of each stritch.

Body Plug. A plug which is used for sealing a tapped hole in a body boss or drain boss.

Body Seat Facing. Refers to a deposit on the body seat ring (or body) of different material on which the body seat is machined.

Body Seat Ring. Refers to the part of a renewable seated valve on which the body seat gets machined. It is made separate from the body and secured in it.

Bogie (bogie truck). Refers to a truck of short wheelbase resting on two or more pairs of wheels, which forms a pivoted support at one or both ends of a long vehicle like a locomotive or a railway coach. Its use has been to enable a long vehicle to run round sharp curves.

Bogie Frame. The frame upon which the axles of a bogie have been mounted.

Boiler. Refers to a steam generator of one or two main types, (*a*) and (*b*)

(*a*) Refers to a shell boiler, in which the water has been contained within more or less cylindrical vessels traversed by tubes, through which the heated gases of combustion pass to impart their heat to the water, like in a locomotive and a bathroom geyser.

(*b*) Refers to a water-tube boiler in which the water is present in the tubes and the heated gases circulate round the tubes on courses directed by baffles.

(*c*) A small water heater which is heated directed by electricity or gas.

(*d*) A flash boiler is having a long coiled tube heated by oil burners which makes water to evaporate as it is pumped through by a feed pump.

Boiler Capacity. Refers to the weight of steam in pounds per hour which a boiler can evaporate when steaming at maximum output.

Boiler Pressure. The term used for the pressure of the steam in a boiler varying according to the type of boiler from a little over atmosphere pressure to more than 10 MPa (1500 lb/in^2) for use with high-pressure turbines.

Boiler (stay) Tap. Refers to a threading tap which is used to tap holes for the reception of boiler stays.

Bolster

    (*a*)   A steel block which is supporting the pad or die in a pressing or punching machine.

    (*b*)   Refers to the bearings that fit within the housings in forge and mill rolls and sustain these rolls.

Bolt. Any piece of material which is used to connect parts together which is having a thread on one end and a head on the other. The head can be hexagonal, square, slotted etc.

    Coach bolt. A bolt which is used to fastern metal parts to wood.

    Eye bolt. A bolt with an eye at the end instead of, or as well as, a head, the latter enabling a degree of side load to be taken by the bolt.

    Rag bolt (foundation bolt). A bolt with a broad, rough tapered bead which is used for attaching machinery or metal structures to masonary, etc.

    Rawbolt (anchor bolt). A proprietary bolt with a surrounding segmented shell which is expanded as a square pyramid shaped nut gets traversed along the bolt thread towards its head during tightening.

    T-bolt. A bolt having its head flush to the shank in one direction and protruding to form a T in the other.

    Unified bolt. A bolt having a unified screw thread.

Bolt Cutter. A machine which is used for cutting the heads of bolts.

Bolt-making Machine (bolt machine). A machine which is used for forging bolts by forming a head on a round bar.

Bonnet (cover)

    (*a*)   Refers to that part of a stop or gate valve which gets attched to the body and carries the operating mechanism; in general, a movable protecting cover and hence 'bonneted safety valve'.

    (*b*)   A cover over an engine as in a motor-vehicle.

Book-folding Machine (or folding machine). A machine used for folding sheets, and gathering, sewing and binding them, which gets adapted to fold sheets of various sizes from folio downwards.

Boom. The jib of a derrick crane.

Boost. Refers to the increase above atmospheric pressure of the induction pressure of a supercharged piston engine (usual units; Pa or lb/in^2).

Boost Gauge. An instrument which is used for measuring the manifold pressure of a supercharged piston engine either in relation to ambient pressure or in absolute terms.

Booster Pump

(*a*) A pump which is used for increasing the pressure of a liquid in a closed pipe system.

(*b*) A pump which is used for maintaining a positive pressure between the fuel tank and an aero-engine of liquid propellant rocket engine.

Bore

(*a*) Refers to the internal diameter of a pipe or a cylinder.

(*b*) Refers to the interior of a small-arm or piece of ordinance, including both the chamber and the rifled portion.

(*c*) Refers to the internal wall of an engine or pump cylinder.

Boring

(*a*) Refers to the operation of machining a cylindrical hole in or through any piece of work, done in a lathe or boring mill. Boring usually presupposes an existing hole which is to be made true and enlarged to the proper size.

(*b*) Refers to the process of drilling holes into the ground or rock for the insertion of blasting charges or to obtain information about the ground or rock.

Boring Bar. A stiff cylindrical bar which is supported at the machine table, carrying the boring tool and driven by the spindle of a boring machine or held in the toolpost of a lathe. Various types have been shown in Figure.

Boring Head (cutter head). Refers to the ring carring the cutters of a boring bar.

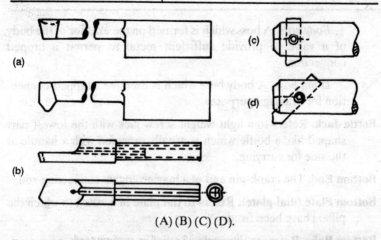

(A) (B) (C) (D).

Boring Machine. A machine tool which is used in boring cylinders the work being clamped on a bed and tool carried by a driving spindle and rotating in fixed supports. The 'horizontal' boring machine has been designed either (*a*) with a spindle which only rotates, or (*b*) with a spindle which both rotates and has a horizontal movement. The 'vertical' boring machine is having a vertical spindle, is similar to a radial driller and for boring cylinder-blocks is having the requisite number of spindles to finish the operation in one pass of the tools.

Boring Mill. Refers to a vertical boring machine in which the work is done by the rotating table and the boring bar is fixed.

Boring Stem. Refers to a heavy bar on which the drill bit gets mounted in boring artesian wells and which adds force to the blow by its weight.

Boring Tool. A tool which is used for internal turning and held on a boring bar. It is having usually only a single cutting edge whereas a drilling tool is having two edges placed on opposite sides of the axis of the tool.

Boss

   (*a*)   Refers to the centre or hub of a wheel.

   (*b*)   Refers to the large part of a shaft on which a wheel has been keyed or at the end where it has been coupled to another shaft.

Body boss. A boss which is formed on the exterior of the body of a valve to provide sufficient metal to permit a tapped connection.

Drain boss. A body boss which is used for a tapped connection for drainage purposes.

Bottle Jack. Refers to a light-weight screw jack with the lowest part shaped like a bottle which is usually provided with a handle at the side for carrying.

Bottom End. The crank-pin end of a marine engine connecting-rod.

Bottom Plate (dial plate). Refers to the plate in a watch to which the pillars have been fixed.

Bottom Rake. Refers to the angle of relief in cutting tools.

Bottom Tool. Refers to the lower half of a fullering tool.

Bottoming. Impinging against the bottom so as to impede free mechanical movement as when a cog strikes the bottom of a space between two cogs.

Bouncing Pin Detonation Meter. An apparatus which is used for fuel testing by determining quantitatively the degree of detonation in a piston engine.

Boundary Conditions. The values of stress, displacement or slope at the ends or edges of a member of finite element in an evaluation of the strength of the item.

Boundary Lubrication. Refers to lubrication by a very thin closely adherent film of oil between two surfaces, the oil, such as a long-chain fatty acid, being one, or at the most, only a few molecules thick.

Bourdon Gauge (pressure gauge). It is a metal tube of a flattened oval section, which gets bent to a curve, with the free end closed and the fixed end open to the pressure. The pressure tends to straighten the tube and the movement is recorded on a dial. Hence sometimes termed as 'dial gauge'.

Bow.

 (*a*) Refers to the ring of a pocket-watch case to which the watch chain gets attached.

Box 45

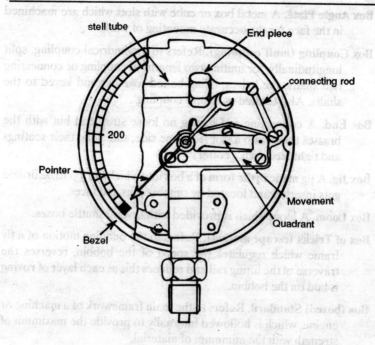

stell tube
End piece
connecting rod
200
Pointer
Movement
Quadrant
Bezel

Bourden pressure gauge.

   (b)   Refers to a flexible strip of whaleborne or cane with the ends drawn together to give tension to a line which has been given a single or pair of turns round a pulley to form a sensitive drive for a drill or mandrel. Hence 'bow drill'.

Bow Connecting-rod (Banjo Frame, Kite Connecting-rod). Refers to a triangular connecting-rod which finds use in steam-pumps with the crank driving the fly wheel enclosed by the bow.

Bow Drill. A small drill which is rotated by hand via the frictional grip of a spring held in tension by a bow.

Bow Saw. Refers to a thin-bladed saw which is held in tension by a special frame or sometimes by a bow.

Box

   (a)   Refers to a portion of a mechanism resembling a box such as a valve box.

   (b)   Refers to a bearing for a shaft.

Box Angle Plate. A metal box or cube with slots which are machined in the face for the accurate mounting of work.

Box Coupling (muff coupling). Refers to a cylindrical coupling, split longitudinally, for uniting two lengths of coupling or connecting two shafts, the halves, being bolted together and keyed to the shafts. Also termed as 'sleeve coupling'.

Box End. A connecting-rod having no loose strap end but with the brasses thrust into a slot from one side, slid along their seatings and tightened with a cotter.

Box Jig. A jig made in the form of a box into which the job to be drilled gets inserted and located by suitable pins and faces.

Box Loom. A loom which is provided with several shuttle boxes.

Box of Tricks (escape motion). Refers to the building motion of a fly frame which regulates the speed of the bobbin, reverses the traverse of the lifting rail and reduces this as each layer of roving is laid on the bobbin.

Box (boxed) Standard. Refers to the main framework of a machine or engine, which is hollowed internally to provide the maximum of strength with the minimum of material.

Box Tools. Combinations of separate tools which are secured in a box for attachment to the faces of a lathe turret.

Bracket. Refers to a support for a machine part.

Brake. The term used for device or mechanism which is used for applying frictional resistance to the motion of a body and thereby absorbing mechanical energy by transferring it into heat (*a*) to retard a vehicle, or (*b*) to measure the power developed by an engine or motor.

Brake Band. The strap or band which is encircling a brake drum.

Brake Block. Blocks of material which are used for applying a frictional force in a brake.

Brake Drum

    (*a*)   Refers to a drum or pulley attached to a wheel (or shaft), to which is applied an external band or internal brake shoes.

    (*b*)   A large drum which is used for winding the rope which lifts and lowers the cage in a colliery or pulls the trucks in

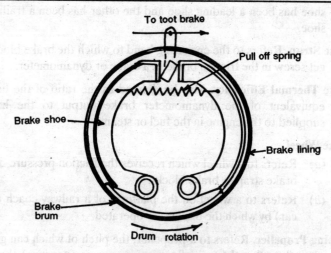

To toot brake

Pull off spring

Brake shoe

Brake lining

Brake brum

Drum rotation

Internal expanding brake.

quarries or operates the lifts in buildings. Also termed as a 'winding drum'.

Brake Horse-power (BHP). Refers to the horse-power which has been developed by an engine as measured by a brake or dynamometer applied to the driving shaft.

Brake Lining. Asbestos-based fabric which is riveted or bounded to the shoes of internal expanding brakes for increasing the friction between them and the drum and, at the same time, to provide a renewable surface.

Brake Mean Effective Pressure (BMEP). Refers to that part of the indicated mean effective pressure (i.m.e.p.) which would provide an output equal to the brake horse-power of an engine or engine cylinder; the product of the i.m.e.p., and the mechanical efficiency.

Brake Power. Refers to the frictional resistance developed by a brake.

Brake Press. Refers to a press in which the energy of a large fly wheel is suddenly applied for forming and blanking.

Brake Shoes. Refers to the internal expanding members in a brake drum on which the renewable friction linings have been mounted. The two shoes in the brake drum of an automobile may be arranged as shown in Figure in this arrangement one shoe has been a leading shoe and the other has been a trailing shoe.

Brake Strap. Refers to the encircling band to which the brake blocks get screw in the friction brake of a crane or dynamometer.

Brake Thermal Efficiency. May be defined as the ratio of the heat equivalent of the dynamometer brake output to the heat supplied to the engine in the fuel or steam.

Brake Wheel

(*a*) Refers to a wheel which receives the friction pressure of a brake strap or brake blocks.

(*b*) Refers to a wheel on the platform of a railway coach (or car) by which the brakes are operated.

Braking Propeller. Refers to a propeller, the pitch of which can gets altered so that the propeller gives a reverse (negative) thrust and acts as an airbrake or waterbrake.

Bramah's Press. The hydrostatic press. ·

Branch Chuck. A chuck which is having four branches turned up at the ends and each furnished with a screw to grip the work.

Brass Finisher's Lathe. A lathe which is specially designed with attachments to machine brasswork in quantities.

Brass Tool. A tool which is used for machining purposes on brass or bronze, usually with no top rake, the front rake and side clearance being about $6°$ with a side rake of 0 to $3°$; drills have straight flutes.

Brass Winding. Filling brass bobbins with lace thread collectively from a jack of wood bobbins.

Brazing. The term used for the process of joining two pieces of metal by fusing a layer of spelter or of a brass alloy between the adjoining surfaces without meeting the parent metal of either piece. The filler material is usually drawn by capillary action into the space between the closely adjacent parts to be joined at about $900°$ C ($1652°F$).

Brazing wire. It is a soft brass wire of small gauge. It is used for binding round joint which are to be brazed. Spinkled with borax the wire melts on heating and runs in.

Break Rolls. Refers to the first series of rollers in a flour mill which break up the gain.

Breakdown Crane (accident crane). Refers to balance crane of the portable jib crane type mounted on a motor lorry or railway truck.

Breaking Joint. Refers to a stepping of consecutive joints so that they are not in line, such as the joints of piston rings, plates in shipbuilding etc.

Breaking Pieces

(*a*) Refers to easily replaceable members of a machine which, are made weaker than the remainder and break first under overload and so protect the machine from extensive damage.

(*b*) Refer to shorts lengths of shafting which are used for coupling an engine with the bottom rolls of a rolling mill, or the rolls from one another. These break first under overload.

(*c*) Refers to the weak link connection between an aero-engine and its propeller in case of a crash.

Breast Wheel. Refers to a water wheel in which the water is entering the buckets at about the height of the wheel centre, above termed as 'high breast' and below termed as 'low breast'.

Breech Block. Refers to the steel block which closes the rear of the bore of a breech-loading fire-arm against the force of the charge.

Breech Mechanism. Refers to the mechanism by which the breech of a gun or a fire-arm is closed before firing.

Breguet Spring. Refers to a balance spring of a watch which is having the outer coil raised above the plane of the spiral and the end of the spring bent to a special form before entering the stud. Hence 'breguet-sprung'.

Bridge

(*a*) Refers to the exterior part of an outside screw valve, connected by pillar to the bonnet, in which the actuating thread of the stem engages either directly or through a bush or through a sleeve.

(b) Refers to an arched guide casting attached to the cover of
     a lift or force pump. The free end of the piston or plunger
     rod travels through the central boss of the casting.

(c) Refers to a raised platform or support in a watch or clock,
     usually with two feet, forming a bearing for one of more
     pivots.

Bridge Gauge. A measuring device which is used to detect the relative
movement of two machine parts due to wear.

Bridge Piece. Refers to the loose piece of the bed in a gap lathe which
fits and bridges the gap.

Bridge Tree

(a) Refers to the cross-bar of a turbine frame above the casing
     which affords a bearing and central support for the spindle
     in millwrighting.

(b) Refers to a lever for sustaining the foodsteps of a millstone
     spindle, which, when raised or lowered, changes the
     distance between the faces of two stones.

Bridge

(a) Refers to the flange of a slide valve of a steam-engine to
     hold the rod in position.

(b) Refers to the loop forged on a slide-valve rod to embrace
     the back of the valve.

(c) A loop or clip for holding test-pieces in a testing machine.

Brinell Hardness Test. Refers to the measurement of the hardness of
a material by the area of the indentation, after equilibrium gets
reached (about 20 secs), produced by a hard steel ball under
specified conditions of loading. The hardness number has been
the ratio of the load to the curved area of indentation.

Brinelling. Refers to the hardening of the surface of a metal by cold
working. The most frequent of occurrence unwanted brinelling
has been in bearings where highly loaded balls or a mating
surface passes over a fixed point causing hardening or even
ultimate enbrittlement and cracking of the surface.

Brittle Fracture. Refers to a tensile failure without significant plastic
deformation of an ordinarily ductile material.

Brittle Lacquer Technique. Refers to a qualitative method in which the part to be tested is coated by a brittle lacquer which cracks, when the part has been subjected to stress.

Brittleness. Refers to the fracture of a material under low stress and without appreciable deformation.

Broach

(*a*)   Refers to a tapered steel shoft having numerous transverse cutting edges increasing in height along its length. It is used as a tool for smoothing and enlarging holes in metal, or as a smooth tool without cutting edges for burnishing pivot holds in watches.

(*b*)   A straight tool with file teeth made of steel to be pressed through irregular-shaped holds in metal, that cannot be dressed by revolving tools, and for splined shaft fittings.

(*c*)   The pin in a lock which enters the barrel of a key.

A circular pull broach is depicted in Figure 15.

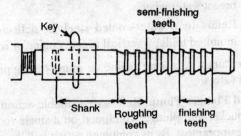

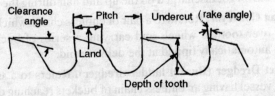

Fig. 15. Broach.

Broaching. Refers to the enlarging, smoothing and turning of holes with a broach or reamer. Broaching is performed on manually-operated presses, on pull-screw machines, on drilling machines, on lathes and on hydraulically-actuated broaching machines or presses. Push broaching is done on vertical machines with short stiff broaches.

Broaching Machine. A machine which is used for finishing square and polygonal holes.

Broad Gauge

(*a*) A railway gauge of 7 ft as laid down by Brunel.

(*b*) A railway gauge greater than the standard width of $8\frac{1}{2}$ in adopted by Stephenson. (7 fit = 2.1336 m approx.; 4 ft $8\frac{1}{2}$ in = 1.4351 m approx).

Bucket

(*a*) Refers to the piston of a reciprocating pump.

(*b*) A cup shaped vane which gets divided midway by a ridge and attached to the periphery of a Pelton wheel.

(*c*) Refers to the cup-shaped receptacle on the impulse wheel of a turbine.

(*d*) Refers to the receptacles for the water in overshot and breast water wheels.

(*e*) Refers to a water-cooled steel jet deflector which gets mounted under a vertical rocket engine or rocket motor.

Bucket Air Pump. A marine engine air pump having piston, foot and head valves.

Bucket and Plunger Pump. Refers to a double-action pump having the bucket and plunger combined on a single rod, the plunger being uppermost. By its combined action half the contents of the barrel got discharged on the up and half during the down stroke.

Bucket Conveyor. Refers to a pair of endless chains which are running over toothed wheels and carrying a series of buckets that are automatically tipped at the delivery end.

Bucket Dredger (Bucket-ladder dredger). Refers to a small-draught vessel having an endless chain of buckets reaching down into the material to be dredged and lifting it for discharge into the vessel or an attendant barge.

Bucket Ladder Excavator. Refers to a mechanical excavator having an endless chain of buckets adapted for excavating on land and discharging into a vehicle for removal.

Bucket Pump. A pump having a bucket or piston which is having valves through it for passage of the fluid lifted.

Bucket Valve. A flap, non-return, valve which gets fitted in the bucket or piston of some types of reciprocating lift pumps.

Buff. A revolving disc which is made of layers of cloth charged with buffer powder for polishing, especially metals.

Buffer Box. Refers to the casing which encloses the buffer spring and buffer rod on rail-mounted vehicles.

Buffer Disc. Refers to the spheroidal disc against which the buffers of rail-mounted vehicles make contact.

Buffer Rod. Refers to the rod carrying the buffer disc.

Buffer Spring. Refers to the spring enclosed in the buffer box which deadens the impact of collision. It is a type of helical spring.

Buffers

(a) Spring-loaded contrivances which are at the ends of railway vehicles to minimize the shock of collision.

(b) Any resilient pad or mechanism which it used for a similar purpose.

Buffing. Refers to the process of polishing as with a buff. Hence 'buffing lathe'.

Building Motion. Refers to a mechanism in fly frames and spinning machines which guides the roving or yarn and builds it into a package.

Building Mover. The term used for a heavy truck on rollers or wide-track wheels. The building rests on a cross-bolster which has been supported by two trucks with at least three rollers each.

Bulger Ram. A round-ended ram which is used for forcing metal plates into apertures, in experiments on bulging stress.

Bulk Modulus (of elasticity). Refers to the ratio of compressive or tensile stress, which is equal in each of three mutualy perpendicular directions relative to the change it produces in volume.

Bull Rope. An endless rope which is able to drive the bull wheel of a cable-drilling ring. The rope gets slipped off the grooved pulley when it is not actually raising or lowering the 'string of tools'.

Bull Wheel

(a)   A large wheel which is engaging with and driving the rack of a planning machine.

(b)   Refers to the driving pulley for the camshaft of a stamp.

(c)   Also refers to the large wheel at the base of a revolving derrick.

(d)   Also refers to the driving wheel of a cable- drilling rig.

Bulldozer

(a)   Refers to a heavy motor-driven vehicle which is mounted on eaterpillar tracks and pushing a broad steel blade in front to remove obstacles, to level uneven surfaces, etc. Also called 'angledozer'.

(b)   Refers to a heavy power-press, which is having a horizontal reciprocating ram for shaping angle irons, etc., with suitable dies. It is used in railway and wagon shops.

Bullck (or horse) Gear. The term used for a mechanism for using animal power by means of a lever attached to gears, the animal walking in a circle.

Bumpers. Refes to the fenders on motor-vehicles, ships etc., for mitigating collisions. They are sometimes sprung when fitted on vehicles.

Bundling Machine. A machine which is used for grasping a number of articles into a bundle ready for tying.

Bung. A stopper which is usually made of a fairly flexible material. It is pressed into a hole to keep out fluids or dirt.

Burnishing

(a)   Spinning.

(b)   Refers to the operation of producing a brilliant finish on metal parts, the edges of books and the surfaces of pottery ware.

Burr

(a) A roughness which is left on a metal by a cutting tool.

(b) Refers to the turned-up edge of metal after punching or drilling.

(c) Refers to a blank punched from sheet-metal.

(d) Refes to a small milling cutter used for engraving and dental work.

(e) Refers to a toothed drum used on a mandrel between lathe centres.

Bursting

(a) The term used for the breaking of a rotating part of machinery due to centrifugal forces.

(b) Refers to the bursting of a vessel due to an excessive pressure difference between inside and outside, the inside being the greater. In the reverse case it is termed as collapsing or bursting inwards.

Bush. A cylindrical sleeve which forms a bearing surface for a shaft or pin. It is usually as a lining. It is having two diameters and the cylindrical length is generally greater than the larger diameter.

Bushing. Means the fitting or driving in of a bush into its seating.

Butt Coupling. Refers to a box coupling for connecting shafting and keyed around the two shafts which butt against each other and are co-axial.

Butt Joint. Refers to a joint between two plates in end contact with a narrow strip riveted or welded to them.

Butterfly throttle

(a) An elliptical plate which is pivoted on its centre. It throttles the steam passing into a cylinder or closes the induction pipe completely.

(b) A circular plate which is used to control the volume of air, and therefore the air/petrol mixture entering a petrol engine.

Butterfly Valve

(a) A pair of semi-circular plates which are hinged axially to a common diametral spindle in a pipe so that the plates allowed flow in one direction only.

(*b*)   A disc which is acting as a throttle when turned on its
diametral axis in a pipe.

Button-headed Screws (half-round screws). Screws having hemi-
spherical heads, slotted for a screwdriver.

Buttress Screw Thread (leaning thread). A screw thread which has
been designed to resist heavy axial loads with the front or thrust
face perpendicular to the axis and the back of the thread sloping
45°.

Bypass. A passage through which a gas or liquid is allowed to flow
instead of or additional to, its ordinary channel or any device for
arranging this.

Bypass Engine. A turbojet engine which is having a relatively large
low-pressure compressor and by-passes some of the air from it
round both the high-pressure compressor and the turbine into
the tail pipe.

Bypass Valve. A valve which is used for directing flow through a
bypass.

Cal. Abbreviation of Calorie.

CAD (computer aided design). Computer Aided Design which is able to describe any or all of a multitude of graphic, computational and engineering design functions done with the aid of a computer and resulting in the choice of a final design. These range from the aesthetic to the most detailed analytical computations.

CADCAM. Abbreviation of Computer Aided Design Combined with Computer Aided Manufacturing.

CADMAT. Abbreviation of Computer Aided Design Manufacture and Test.

CAE. Abbreviation of Computer Aided Engineering.

CAR. Abbreviation of Computer Aided Repair.

CAT. Abbreviation of Computer Aided Testing.

Digital instrument. An instrument displaying its output to the nearest whole number (digit) for each full number of decimal position of the value being disclosed.

Analogue instrument. An instrument displaying the continuous variation in the magnitude of the reading being disclosed.

Digital computer. A computer which is able to do simple calculations at very high speed by electrical impulses related to digist in a code. Very large and complex calculations can be carried out by reduction of the problems to very laborious but simple solutions.

Analogue computer. A computer which is operating with the continuously changing amplitudes of its vairables. The analogue computer family includes all devices in which measurable physical quantities (usually electrical) are made to follow the mathematical relationships existing in some particular problem.

CEng. Designatory letters for a person who has been a chartered engineer.

Cf. Abbreviation of Centrifugal force.

Cg. Abbreviation of Centre of gravity.

CGS System. Abbreviation of Centimetre-gramme-second system of units.

Cmps. Abbreviation of Centimetres per second.

Compr. Abbreviation of Compression ratio.

CP Cycle. Abbreviation of Constant pressure cycle.

C to C. Centre to centre.

C-spanner. Refers to a flat metal spanner having semi-circular end, the tip of which protrudes inwards to engage in the slot of a slotted ring nut.

C/s, Cps. Cycles per second.

CV Cycle. Constant volume cycle.

Cabinet Leg. The cupboard under a lathe or machine too which is used for holding gear, tools, oil tank etc.

Cable-drilling Rig. Refers to a large earth-drilling machine in which the drill bit at the bottom of a long tube gets suspended on a cable from a vertical derrick and rotated at ground level by a prime mover; the tube can be lengthened by additions at ground level.

Cable Grip. It is a flexible cone of wire put on the end of a cable to make the cable to be pulled into a duct as tension tightens the cone on the cable.

Cable Ploughing. The term used for the ploughing of a field by using two traction engines, connected by a wire rope attached to a plough, which alternately supply the necessary power to pull the plough across the field.

Cable Railway. Refers to a railway with the motive power coming from a continuous moving cable, overhead or underground, to which the car can get rigidly connected by a clutch device at any point on the cable.

Cable Tramcars. Tramcars which are operated from an underground cable in the same manner as a cable railway.

Cable-way (blondin). The term used for the construction for transporting material by skips suspended from cables which have been slung over and between a series of towards, the skips being raised, lowered or moved to any position along the cables.

Cage

(*a*) It is a platform, with or without framework, which is used for lowering or raising goods, etc. A 'cage' in a mine is able to lower and raise men or wagons in a shaft.

(*b*) The frame in a travelling crane within which a man sits and controls its movements.

Calender

(*a*) A machine in which material is allowed to pass through rollers under pressure to impart the desired finish or to ensure uniform thickness, in a steel mill.

(*b*) A machine which is used for calendering soft materials. It is usually consisting of a number of rollers.

Calender Rollers

(*a*) Heavy grooved rollers which are used for feeding timber into sawing or planing machines.

(*b*) The term used for the rollers in a calender.

Calendering. Refers to the series of operations (varying according to the goods) of straightening, damping, pressing etc., woven goods to provide them the desired finish.

Caliper. The size of a watch movement.

Callipers. They are made like dividers or a compass. They are having curved hinged legs for measuring thickness and outside diameters, outside callipers, (points inward) or inside measurements, inside callipers (points outward).

Jenny Callipers are having one point straight leg and the other curved inwards to make measurement from a point to a convex surface.

Great-tooth callipers are having both of their ends pointing in the same circumferential direction and can be used for measuring the normal pitch from which the circular pitch can be calculated for a known diameter.

Calliper Gauge. Refers to a horseshoe type of limit gauge having two pairs of jaws marked 'Go' and 'No go' corresponding to the tolerance allowed on the dimensions for the work.

Callipers, Poising. Refers to a form of cross-over callipers with jaws between which a balance for a watch can get mounted and rotated to test for truth and poise, that is, static and dynamic balance.

Press to open

Cross over callipers adapted for poising.

Cam. Refers to a shaped component of a mechanism, like a heart-shaped disc on a shaft, which decides the motion of a follower. Cams are generally described by their shape such as Tangent Cam or Circular Arc Cam. The motion of the cam could be linear or rotary, usually the latter.

Barrel cam. It is a cylindrical cam having a track for the cam follower on its end or its circumference.

Bi-polar cam. It is a pair of shaped plates with the teeth which engage so that the input and output shafts of the plates all having a varying velocity ratio.

Globoidal cam. It is a cam of cylindrical type as in Fig. (c) but where the cylindrical is not of constant diameter. It is having a globoid profile, being of smaller diameter at the centre.

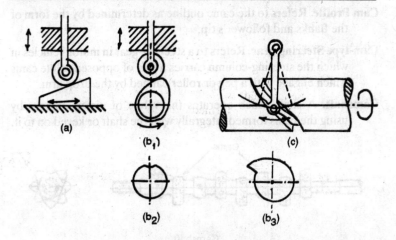

Cams (a) Wedge type (b) Radical or Disc Type (c) Cylindrical.

Pinwheel cam. See pinwheel gear.

Plate cam. A cam which is profiled out of a plate material.

Radial cam. This term is often further described by the action it produces, such as constant velocity, constant acceleration/declaration or simple harmonic motion.

Cam-ball Valve. A valve which is actuated by a cam on the axis of a ball-lever, the ball rising with the level of water in a cistern, tank or boiler, so that the cam makes the valve of shut off the supply.

Cam Chuck. A profiling device which is fitted to the saddle of a lathe for turning irregular forms such as cams. The cutter turns in the fixed headstock and the work gets manipulated on the rest.

Cam Follower. Refers to that part of an engine or mechanism which rides on the contour surface of a cam and to which motion has been imparted by the cam.

Cam Governor. A stepped or differential cam, giving three or four grades of lift, has its action controlled by governor balls which slide a roller on to one on another of the cams according to the centrifugal action on the balls; it is found in Otto-cycle gas engines.

Cam Lobe. Refers to the raised portions on the counter of a cam which operate the cam followers.

Cam Profile. Refers to the came outline as determined by the form of the flanks and follower's tip.

Cam-type Steering-gear. Refers to a steering-gear in motor vehicles in which the steering-column carries a pair of opposed valute cams which engage with a peg or roller carried by the drop arm.

Camshaft. A shaft which operates the valves of piston engines by using the cams formed integrally with the shaft or keyed on to it.

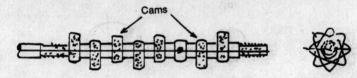

Camshaft.

Cancelling Machine. A rotatory cylinder having a ratchet which is used in post offices for cancelling the stamps on letters.

Cannon. A hollow spindle or shaft with a motion independent of an internal spindle or shaft.

Cannon Pinion. Refers to the pinion with an extended pipe to which the minute hand of a watch is usually attached.

Cannon Wheel. Refers to the wheel on the centre arbor which is carrying the minute hand on a square section of its pipe.

Cannon Wheel Spring. A small plate which is bent upwards from the front plate and acting as a spring washer for the cannon wheel.

Canting. Refers to the tilting over of the moving part of a mechanism from its proper angle when in motion or tilting machinery at an angle for a special purpose such as for cutting a bevel or for cutting at a definite angle.

Cap Jewel. A jewel having endstone in horology.

Cap Screw. A screw-bolt having a hexagon head used without a nut, being attached to a threaded hole in the adjacent part of the assembly.

Capping. Refers to the shrouding of gear wheels.

Caprotti Valve-gear. Refers to two pairs of vertical double-beat poppet valves which get operated by cams with an adjustable cut-off obtained by varying the angular position of the inlet cams, as found on some locomotives.

Capstan. Refers to a vertical cone-shaped drum or spindle, on which a rope or chain gets wound, that is rotated by man, steam, hydraulic or electric power; for example, for warping a ship or hoisting an anchor.

Capstan Engine. A capstan driven by steam power, the general arrangement resembling a winch with worm gearing operating the vertical drums.

Capstan-headed Screw. A screw having a cylindrical head pierced by radial holes so that it can be tightened by a tommy bar inserted in the holes.

Capstan Lathe. A lathe in which the tools get mounted in a capstan tool head.

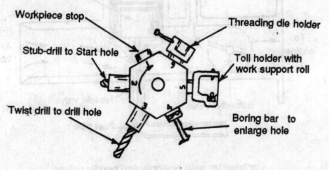

Workpiece stop

Stub-drill to Start hole

Twist drill to drill hole

Threading die holder

Toll holder with work support roll

Boring bar to enlarge hole

Capstan tool head.

Capstan Tool Head. Refers to the support of the hexagonal tool-past of a capstan lathe. It gets mounted on a short slide which in turn is part of a carriage sliding on the lathe bed. The capstan rest is actuated by the star wheel. This construction provides a short working stroke allowing rapid manipulation of the hexagonal toolpost.

Carbon Gland. Refers to type of gland in the form of segmented (arbon) rings to disallow leakage along a shaft as in steam-turbines and on shafts driven from outside chambers full of air at high pressure.

Carburation. Refers to the mixing of air and fuel to make a combustible mixture using carburetter.

Carburetter (carburettor). Refers to a device in which a fuel gets atomized and mixed with air.

The pressure difference across a piston caused by the air velocity though the nozzle inlet (choke) regulates the fule flow by using a needle valve attach to the piston. The variable demand to an engine is obtained by varying the flow into the engine through a butterfly valve. A simplified illustration is shown in Figure.

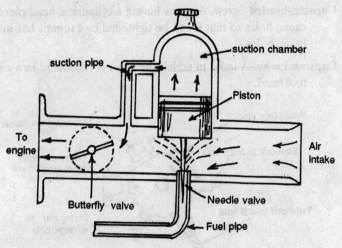

Carburetter with butterfly and needle valves.

Cardan Shaft. A shaft which is transmitting power as in a motor-vehicle or the propeller shaft in a ship.

Carnot Cycle. Refers to the working cycle of an ideal heat-engine of maximum thermal efficiency, consisting of isothermal expansion, adiabatic expansion, isothermal compression and adiabatic compression to the initial state.

Carnot's Law. No engine has been more efficient than a reversible engine working between the same temperatures. The engine's efficiency has been independent of the nature of the working substance, being dependent on the temperature only.

Carriage

(*a*) Refers to that part of a lathe which slides on the lathe bed and carries the cutting tool.

(*b*) Refers to the horizontal table of a printing machine which travels to and for under the cylinder or roller.

(*c*) Refers to that part of a lace machine carrying the bobbin thread and swinging in an arc on the combs which contain the warp threads.

Carriage Clock. A small portable clock with a platform escapement, usually having a brass case and glass panels. Sometimes called 'travelling clock'.

Carriage Spring. Refers to any elastic device, often curved steel strips of varying length, interposed between the bed of a vehicle and its running gear.

Carrier

(*a*) Refers to a receptacle used in connection with a conveyor system.

(*b*) Refers to a device screwed to the work and driven by a pin projecting from the faceplate of a lathe, so that the work is rotated between centres.

(*c*) Refers to the first roller in a carding engine which unwinds the lap and distributes it to the machine.

(*d*) Refers to an intermediate roller on a coarse wool-carder (scribbling machine) between the feel rolls and the toothed drum.

Carryall. A self-loading, self-discharging transport vehicle, either towed or with a built-in power unit.

Caster (castor). Refers to a small wheel on a swivel.

Caster Angle (wheel action). Refers t the inclination of the king-pin in a motor-vehicle in such a v .y that its axis intersects the ground at a point in advance c' the point of contact of the wheel with the ground, thereoy yielding a self- centring tendency to the steerable front wheels after angular deflection by road shocks.

Casting

(*a*) A metallic object which is produced by casting as distinct from being machined or formed by any other means.

Sand casting. A pattern is embedded in sand and then removed to leave a cavity of the needed shape and size. Metal is then poured into the cavity where it cools to form a casting.

Die casting (diecasting). Refers to the casting of metals or plastics into permanent metal moulds. Aluminium, zinc, copper, tin, and lead-base alloys have been suitable for this purpose.

Centrifugal casting. Cylindrical and sherical components can be cast in a rotating mould so that the centrifugal force produces a fine grain high-quality casting.

A metal or plastic article which is formed by pouring liquified material into a mould as distinct from one shaped by working on a blank.

Castle Nut. Refers to hexagonal nut with six radial slots, any two of which can line up with a hole drilled in the bolt or screw for the insertion of a split pin to disallow loosening.

Cataract. Refers to a kind of hydraulic brake which consist of a plunger and valves for regulating the action of a pumping engine and other machines.

Cataract rod. Refers to a vertical rod attached to the lever of a cataract which directly operates the above action.

Catch Bar. Refers to a long bar of steel forming part of a lace machine and faced with brass on the part, called a 'driving blade', which engages with the carriages.

Catch Plate. Refers to the end flange of a lathe-sheed cone, or of an internal plate driven by the cone through a hole in which a removable peg is taking drive to the lathe mandrel.

Caterpillar (track). Road wheels replaced by endless articulated steel bands of flate plates, or by chains which are passing round two more wheels to enable a vehicle to cope with rough and uneven ground. Removable projecting pieces have been some- times provided to increase grip on soft ground; the caterpillar also spreads the load.

Cathead (Spider). A turned sleeve which is having four or more radial screws in each end. It is used on a lathe for clamping on to rough work of small diameter and running in the steady while centring.

Centre Distance. Refers to length of a common perpendicular to the axes of two gears in mesh.

Centre Drill. A small drill with straight flutes. It is used for drilling holes in the end of a bar. It is mounted between centres in a machine tool; the drill is so shaped that it produces a counter-sunk hole.

Centre Gauge (screw-cutter's gauge). A thin metal gauge with vee-shaped notches which is having an included angle of 55°, around its parimeter. It is used as a template for turning the cone points of lathes, grinding the angles of screw-cutting tools and setting these tool-post.

Centre Lathe (engine lathe). A machine which is used for carrying out turning, boring of screw-cutting operations on work held between centres or in a chuck, but not repetition work.

Centre of Gyration. Refers to the point in a revolving body in which its angular momentum gets concentrated.

Centre Punch. Refers to a steel punch with a conical point used to make, by indentation, the centres of holes to be drilled etc.

Centre Wheel (horology). The wheel mounted on the arbor of the centre pinion; it usually makes one turn per hour.

Centre-weighted Governor. Refers to a high-speed small governor with a heavy weight sliding on a central spindle, whose gravity gets balanced by the centrifugal force of the balls.

Centreless Grinding. Grinding by using a 'centreless grinder' where the work gets supported and fed to the grinding wheel on a knife-edge support, substituting the two fixed centres of conventional cylindrical grinding machines.

Centrifugal Brake. An automatic brake which is actuated by revolving brake shoes forced out by centrifugal force into contact with a fixed brake drum.

Centrifugal Clutch. It is a clutch in which friction surfaces engage at a definite speed with the driving member and get engaged due to the centrifugal force exerted by weighted levers.

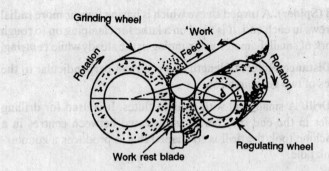

Centreless grinding.

Centrifugal Compressor. Refers to an air compressor in which the pressure rise is obtained by the centrifugal forces set up by a rotating impeller.

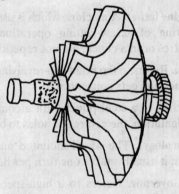

Centrifugal compressor, single entre.

Centrifugal Fan (paddle-wheel fan). It is a fan of paddle-wheel from in which the air entres axially at the centre and gets discharged radially by centrifugal force.

Centrifugal Pump. It is a pump which consists of one or more impellers equipped with vanes mounted on a rotating shaft and enclosed by a casing. The liquid gets drawn into the centres of the rotating impellers and flows out radially under centrifugal force; the resulting kinetic energy would get converted into pressure energy in the casing or diffuser.

Centring Machine. A machine which is used for facing forgings, etc., and marking the centre for subsequent turning.

Centroid of an Area. Centre of gravity of an area.

Ceramic Tool. A cutting tool, which is suitable for machining almost all metals and abrasive materials. It could be dressed only by diamond wheels or, under light pressure, by green grit wheels.

Chain. Metal links of oblong or circular shape which gets interconnected to form a flexible cable to be used for hoisting or for power transmission.

Block chain. A chain which consists of blocks connected together by links and pins and used for comparatively low speeds.

Chain drive. Refers to a series of pairs of links interconnected at their joints by parallel rods. It is often surrounded by hardened steel rollers, which engage with teeth on sprocket wheels to transmit power between two parallel shafts.

Close link chain. It is an open link chain, with the length of links not greater than five times the diameter and with a width of three and a half diameters not to be confused with 'circular link chain'.

Coil chain. It consists of oblong links of circular section usually of welded wrought-iron or steel. The links may be plain or have stud (or bridge) across the centre; the studs tend to prevent stretching and kinking.

Detachable link chain. It consists of detachable and replaceable links. It is used for low-speed and light-load power transmission, and for conveyors and elevators of moderate capacity and length.

Double duplex, dual chain. It is a pair of adjacent chains forming a chain drive of double width. Triple and quadruple chains can be used for transmitting high powers.

Duplex chain. It is a double width roller chain having two sets of rollers and three links on double length pins. These are used to transmit high power.

Pintle chain. A sprocket chain.

Plate link chain. It is a chain with flat links united by pins passing through holes near the ends of the links.

Roller chain. (*a*) It consists of several rollers which are held in links. There are used to form a linear bearing.

(*b*) This chain consists of alternate links held by connecting-pins which are fastened by cotters, the pins serving to carry the rollers which bear on the sprocket teeth.

Silent chain. This chain consists of alternate flate steel links connected by pins.

Sprocket chain. A chain which is suitable for use on a toothed wheel.

Studded chain. A chain having stud links.

Chain Barrel. Refers to a cylindrical barrel, sometimes grooved, on which surplus chain is wound.

Chain Conveyor. A conveyor with endless chains supporting slats, buckets, etc., as distinct from the use of a single band.

Chain Cutter. A cutter composed of the links of an endless chain. It is made of special tool steel and ground with a hook of 25° on the outsides of the links. The pitch has been the length of two links.

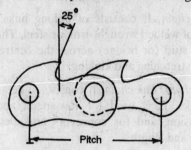

Morties chain cutter.

Chain Gearing. A gearing which uses a chain and wheels in which projections on the wheel fit into cavities in the chain or vice versa.

Chain Saw. A power-driven endless chain having saw-like teeth on its links.

Chamfer. Refers to a bevel produced on edges or corners which are otherwise rectangular.

Chamfering Machine. A machine which is used for forming the bevels or nuts and rounding the ends of bolts.

Change Wheels. Refers to the gear-wheel through which the lead screw of a screw-cutting lathe gets driven from the mandrel; the wheels get changed to vary the reduction ratio.

Charpy Test. Refers to a notched-bar test in which a specimen, notched at the middle and fixed at both ends, gets struck behind the notch by a striker carried on a pendulum. The absorbed energy gets measured by the decrease in height of the swing of the pendulum after fracture.

Chaser (comb tool). A lathe tool which is used for cutting and finishing internal or external screw threads, usually the latter. The edge of the tool has been the counter part of the screw section.

Chasing. Refers to the cutting or finishing of screw threads with a chaser.

Chasing Attachment. A special feed motion which is built into capstan lathes and turret lathes, the special lead screw being driven by a shaft from the feed box thus allowing large-diamater threads to be formed with a chaser. Small-diameter threaded work is usually formed with a diehead.

Chasis. Refers to the base-frame of a vehicle.

Chatter. Refers to the vibration of a blunt, or badly set or insufficiently rigid cutting tool yielding an irregular surface finish on the workpiece.

Check Gauge. A gauge which is used for checking the accuracy of other gauges, normally for the verification of individual dimensions.

Check Valve (non-return valve). It is a valve which prevents reversal of flow by means of a check mechanism. The valve is opened by the flow of fluid and closed either by the weight of the mechanism when the flow ceases, or by a spring or by back pressure. There have been three patterns (*a*) 'horizontal,' with the body ends in line with each other, (*b*) 'vertical' in a vertical

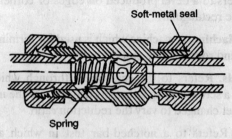

Check valve (non-return valve).

position with the body ends in line with each other, and (c) 'angle' with the body ends at right angles to each other.

Foot check valve (a). A valve which is fitted to the bottom of a suction pipe, usually with a strainer; the lowest valve in a pump.

(b) A non-return valve at the inlet end of a suction pipe.

Lift type check valve. A mechanism incorporating a disc, piston or ball which lifts along an axis line with the axis of the body seat.

Screw-down stop and check valve. A check valve having a mechanism to hold the disc in the closed position independently of the flow or to restrict the lift of the disc.

Swing type check valve. It is a check mechanism incorporating a disc which swings on a hinge.

Cheese Head. Similar to a Fillister head but with a flat top face.

Cheeses. Large bobbins on which have been wound threads of reinforcing fibres used in the production of composite materials.

Cherry Picker (flying cherry picker). A tall mobile platform of variable height which makes an operator to reach otherwise inaccessible parts of a structure.

Chime Barrel. A cylinder having short vertical pins to lift the hammers in a chiming clock.

Chilled Rolls. The rolls which have had their surfaces hardened by chilling.

Chinese Windlass. Refers to two drums or cylinders of slightly different diameter on the same axis with a single coil of rope wound in opposite directions on each. The rope winds off one drum as it has been wound on to the other, giving a slow motion with a considerable mechanical gain and enabling a large weight to be raised very slowly by a small expenditure of power.

Chobert Rivet. It is a form of blind rivet having a tapered bore and tapered pin, so that the pin is able to make a tight structural connection with the outer sleeve in a final installation.

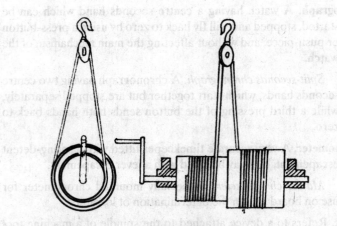

Chinese windlass.

Choke (restrictor)

    (*a*)   A restriction in a pipe to reduce fluid flow.

    (*b*)   Refers to a valve, usually a butterfly valve, in a car-buretter-intake to reduce the air supply and thus give a rich micture for starting purposes or while the engine is still cold. Also called 'strangler'.

    Bleed choke. A choke which is used for releasing fluid, usually to atmospheric pressure.

Chopper

    (*a*)   Refers to any device which interrupts regularly some quantity, such as light into a photocell.

    (*b*)   A term used in the USA for a helicopter.

Chops. Refers to two flat pieces of metal acting as a clamp to secure the end of a pendulum suspension spring.

Chordal Height. Refers to the shortest distance from the mid-point of the chord, which is to be measured, to the tooth crest of a gear: it has been measured on the back cone of a bevel gear.

Normal chordal height. Refers to the shortest distance from the tooth crest to the reference cylinder of a gear.

Chordal Thickness. Refers to the thickness of the gear-tooth measured at the pitch circle.

Chonograph. A watch having a centre seconds hand which can be started, stopped and will fly back to zero by using a press-button or push-piece, and without affecting the main mechanism of the watch.

Split-seconds chronograph. A chronograph having two centre seconds hands, which start together but are stopped separately, while a third pressing of the button sends both hands back to zero.

Chronometer. A very accurate timekeeper fitted with a spring-detent escapement, but may be fitted with a lever escapement.

Marine chronometer. A specially mounted chronometer for use on board ship in the determination of longitude.

Chuck. Refers to a device attached to the spindle of a machine tool which grips the rotating drill, cutting tool or work, such as work secured to the mandrel of a lathe's headstock (See Figure).

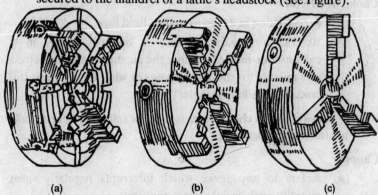

(a)   (b)   (c)

Chucks: (*a*) four-jaw independent, (*b*) four-jaw self-centring,
(*c*) three-jaw self-centring.

Four-jaw self-centring chuck. A chuck having four jaws disposed at right angles but a single key moves all four jaws simultaneously in or out via spiral and mating segments on the back of each jaw.

Independent four-jaw chuck. A chuck having four jaws disposed at right angles to each other and each separately adjustable. It is used for holding rectangular circular or irregular workpieces.

Chucking. Attaching lathe work to a chuck.

Chucking Machine. Refers to a machine tool in which the work has been held and driven by a chuck, but not supported on centres.

Chucking Reamer (straight shank reamer). A reamer having a cylindrical shank for use in a self-centring chuck.

Cinematograph. An apparatus which is used for projecting film and reproducing synchronized sound.

Circlip (retaining ring). It is a clip of a spring steel in the form of an incomnplete ring which fits tightly into a groove of a shaft in the case of an external clip, or into the groove within a bore in the case of an internal clip, for locating a pair of mating parts in an axial direction. A circlip could be of circular or rectangular cross section the latter often having internal or external eyes for easier fitting using special tools like pliers.

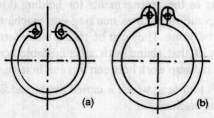

Circhips: (*a*) in (*b*) external.

Circular-form Tool. Refers to a typical circular-form tool which is mounted in its holder.

Circular Pitch (circumferential pitch). Refers to the pitch of wheel teeth which is measured along the circumference of the pitch circle (rolling circle), upon which one wheel comes into contact with its mate.

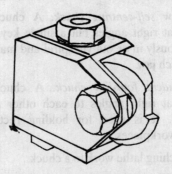

Circular-form tool in holder.

Circular Saw (buzz saw). A steel disc having teeth on its circumference which variety from a few centimetres up to a few metres diameter (or a few inches up to seven feet). It is used for cutting wood, metal and other materials.

Circular Table. Refers to a circular cast-iron plate, which is able to sustain the work in drilling and slotting machines.

Circulating Path. A path having a solvent or acid in which the fluid gets stirred so as to subject all parts of the immersed body to the same strength of fluid.

Circulating Pump. A pump which is used for circulating cooling water through the condenser of a steam plant.

Clamps. Refers to the arrangements for holding down workpieces during operations such as marking-out, machining, measuring, grinding or fitting. They can be taking many forms. A G clamp has been of that shape with an adjustable screw across the opening to clamp work between top and bottom.

Clapper Box. A tool-head which is caried on the saddle of a planning or shaping machine.

Claw Clutch (claw coupling). Refers to a shaft coupling for instant connection or disconnection in which flanges carried by each shaft engage, through teeth in corresponding recesses in their opposing faces, one flange being slidable axially for disengagement.

Claws. Refers to the points or claws that operate on the sprocket holes of a cine film and thus intermitenly feed the film forward through the picture gate for projection.

Clearance

(*a*) Refers to the distance between two objects or between two mating parts when assembled together.

(*b*) Refers to the distance between a moving and a stationary part of a machine or between two moving parts.

(*c*) Refers to the angular baking-off given to a cutting tool so that the heel will clear the work.

Major (minor) clearance. Refers to the distance between the design forms at the root (crest) of an internal screw thread and the crest (root) of the external screw thread.

Clearance Volume. Refers to the volume which is enclosed by the piston and the adjacent end of a reciprocating engine cylinder or of a compressor when the crank has been on the inner dead-centre.

Clearing Hole (clearance hole). A hole which is full to the specified size so that an object, such as a bolt or stud, of the same nominal size will pass through it.

Clerk Cycle. A two-stroke internal-combustion engine cycle which is produced by Donald Clerk 1878-1881 and developed by Joseph Day in 1891 who dispensed with Clerk's second cylinder and employed the crankcase as we know it today.

Clevis Pin. It is a pin or dowel with head having a hole near the tail end of the shaft to take a clevis. A clevis pin has been generally used to connect two or more items together via concentric holes in the items. Withdrawl of the clevis pin is disallowed by insertion of the clevis which is inserted, generally after a plain washer has been slid along the clevis pin.

Click. A pawl used in horology in connection with a ratchet wheel to allow rotation in one direction only.

Click spring. Refers to the spring which holds the click in the teeth of a ratchet wheel.

Click wheel. Refers to a small ratchet wheel.

Climb Milling. Refers to the milling process when the work gets fed in the same direction as the path of the teeth on the cutter. This is the opposite of conventional milling.

Clip Pulley (clip drum). A rope pulley having a rim of vee-section which is constructed of movable clips, about 10 cm (3 - 4 in) long, hinged on pins whose axes are along the direction of the pulley's periphery. The effect of the rope biting has been to pull the clips towards each other and thus increase the bite.

Clips. Metal bands or wire formed into a circle to clamp flexible tubes to solid pipes, or fix cables, pipes and tubes at a chosen spot with the help of a bolt and nut or similar device.

Closing-up. Refers to the operation of forming a head on a rivet shank.

Cluster Mill. A mill which consists of two small working rolls each supported and driven by a pair of two large rolls.

Clutch. Refers to the coupling of two working parts, for example two shafts, in such a way as to allow connection or disconnection at will without the necessity of bringing both parts to rest, and when connected transmit the needed amount of power without slip.

'Friction clutches' operate by surface friction when two surfaces get pressed together. 'Magnetic clutches' use the attraction of a magnet for its armature or compaction of fine metallic particles as in the dry fluid clutch. 'Jaw clutches' give a positive drive by using projecting lugs.

Band clutch. It is a fabric-lined steel band which has been contracted on the periphery of the driving member by means of engaging gear.

Block clutch. Friction shoes which are forced inwards into the grooved rim of the driving member or expanded into contact with the internal surface of a drum.

Centrifugal clutch. It is a friction clutch which gets engaged automatically at a definite speed of the driving member and is maintained in contact by the centrifugal force exerted by weighted levers.

Coil clutch. It is a friction clutch which used a coil of steel around a drum.

Cone clutch. It is a friction clutch in which the internally coned member can get moved axially in or out of the externally coned member for engaging or disengaging the drive. The cone clutch has been either metal-to-metal running in an oil path or fabric-to-metal running dry.

Disc (plate) clutch. It is a friction clutch in which both the driving and the driven members are having flat circular or surfaces that are brought into contact and has one or more annular discs running either dry or lubricated, being called respectively single plate clutch or 'multiple-disc clutch'. The former is generally fabric-faced and the latter usually run in oil; both gets loaded by springs.

Dog clutch. It is a jaw clutch which consists of opposed flanges, or male and female members, provided with projections and slots with one member slidable for engaging and disengaging the drive.

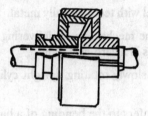

Cone clutch.

Dry fluid clutch. It is a clutch in which metal particles get compacted, under the action of the centrifugal force produced by rotation of the driving member, or by an applied magnetic field.

Split ring clutch. It is a friction clutch which consists of a split-ring which get expanded into a sleeve by a cam or lever mechanism. It is commonly used in machine tools.

Sprag clutch. It is a clutch in which balls or D-shaped pieces get wedged between driver and driven members with rotation in one direction but are free when rotation is in the opposite direction.

Coaxial Propellers. Two propellers which are mounted on an aeroplane on concentric shafts with independent drives and rotating in opposite directions.

Cock

    (*a*)   Refers to a carrier or bracket for a pivot in horology.

    (*b*)   Refers to a tap or cylindrical valve for controlling the flow of a liquid or gas.

    (*c*)   Refers to a lever in a firearm, raised ready to be released by a trigger.

Cocking Lever. A lever which is used for raising the cock or hammer of a gun in readiness for firing.

Cogging

    (*a*)   Refers to the operation of rolling or forging an ingot to reduce it to a bloom or billet.

    (*b*)   Refers to the fitting in, and working of, the cogs of mortise wheels.

Cog Wheel. A wheel with teeth, usually metal.

Coiler. Refers to the mechanism of delivering the silver in coils into the coiler cans in a carding engine.

Coiler Can. It is a slowly rotating upright cylinder for receiving the silver in coils.

Cold Bend Test. Refers to the bending of a bar while cold through a given angle to check on materials ductility.

Coldsaw (cold iron saw). It is a metal-cutting slow-running circular saw which is used for cutting steel bars to length.

Collar. Refers to a rectangular section ring secured to, or integral with a shaft to provide axial location for a bearing or to prevent axial movement of a shaft through a thrust bearing.

Collar Bearing. It is a bearing which is provided with several collars to take the thrust of a shaft or to provide adequate surfaces for lubrication of a vertical shaft.

Collar-headed Screw. It is a screw in which the head is having an integral collar to stop any fluid leakage past the threads.

Collaring. Refers to the wrapping of a rolled bar around the bottom roll of a rolling mill.

Collet

(*a*) Refers to a slotted sleeve, externally coned, which fits into the internally coned more of a lathe mandrel and has been used to grip small circular work or tools; also termed as 'collet chuck'.

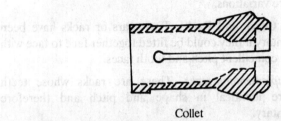

Collet

(*b*) A disc or ring which is used for holding dies or nuts in a screwing machine.

(*c*) A circular flange or collar.

Comb Tool. A chaser.

Combination Chuck. A lathe chuck which can be operated either as an independent chuck or as a self-centring chuck.

Combination Planer. It is a machine in which one cutter-block can be used for surface planning and for thicknessing.

Combination Turbine. A disc-and-drum turbine.

Combination-turret Lathe. A lathe which is capable of automatic turning, facing drilling boring and threading operations by incorporation of an automatic sliding and surfacing saddle and turret.

Combine Baler. It is a machine which gathers hay or straw from the rows in a field and forms it into bales.

Combustion Chamber

(*a*) A chamber in which combustion occurs in an internal-combustion engine, a jet engine or a rocket engine.

(*b*) Any chamber of space in which the combustion of gaseous mixtures or products occurs.

Comparator. An apparatus which is used for the accurate comparison of length standards or for measuring the coefficients of expansion of metal bars or for indicating progress during the manufacture of a part in relation to specified limits.

Compensating Collar. An annular ring which is fitted on a revolving shaft with adjustment to compensate for wear.

Compensation Balance. A balance which has been constructed so as to compensate for dimensional changes and the elastic properties of the balance spring and balance wheel caused by temperature variations.

Complementary Gears and Racks. Two gears or racks have been complementary if they could be fitted together face to face with completely coincident pitch and tooth faces.

Self-complementary racks. These are racks whose teeth profiles are identical in shape and pitch and therefore complementary.

Compliance Mechanism. It is a mechanism which is programmed to include compliance into the kinematics of the end effector so that a defined amount of inaccuracy associated with the object being handled by the robot can get assimilated.

Composite Engine. Refers to a combination of two engines of basically different design, such as a piston-turbine combination.

Compound Engine

(*a*) It is a gas-turbine engine in which the compression is carried out in stages in a number of mechanically sub-divided compressors, each driven by a separate turbine.

(*b*) It is an internal-combustion engine in whose exhaust gases more fuel has been burnt before they pass through a final turbine stage which in turn drives a turbo-charger supplying the main I.C. engine.

Compound Lever. Refers to a series of levers for getting a large mechanical advantage as in large weighing and testing machines.

Compounding. Means expanding steam in two or more stages, either in reciprocating engines or in steam-turbines.

Compressing Cylinder

(*a*) Refers to the cylinder of an air-compressor within which the air is compressed.

(*b*) Refers to a cylinder which is used in some gas-engines for compressing the air in the charge.

Compression. Refers to the total shortening in length produced in a test specimen during a compressive test. It is usually expressed as a percentage of the original length of the specimen.

Compression Coupling. A coupling which connects two smooth shafts of the same diameter with an external longitudinally split double conical sleeve. It is held on to the shafts by the internally tapered interconnected rings compressing the sleeve from either end.

Compression Engine. It is a gas-engine in which the mixed charge gets compressed previous to ignition.

Compression-ignition Engine. It is an internal-combustion engine in which the heat of compression ignites the mixture of air and injected fule in the cylinder.

Compression Ratio

(*a*) Refers to the ratio of the volume of the mixture in the cylinder of a piston-engine before compression to the volume when compressed.

(*b*) Refers to the ratio of the air (gas) pressures across a compressor of a jet-engine.

Compression Test. Refers to the test of a specimen under increasing compressive force often until it fails, and the recording of the stress-strain relationship.

Computer Vision System. A system which is used for robot guidance or information in which optical images and patterns received through a variety of devices have been broken down into digital data and processed by a comuter. These find use for a variety of purposes such part recognition, orientation, etc.

Condenser. It is a chamber into which exhaust steam from a steam-engine or turbine gets condensed; a high degree of vacuum gets maintained by an air-pump.

Cone

(i) The stepped driving pulley used in belting on a machine tool for the governing of different speeds, sometimes called the 'speed cone'.

(ii) The conical race for the balls in certain types of ball-bearing.

Cone Bearing. It is a conical journal which is running in a correspondingly tapered bush, thereby acting as a combined journal and thrust-bearing and used for some lathe spindles.

Cone Clutch. See clutch.

Cone Distance. Refers to the distance from the apex of a bevel gear to the pitch circle. It is measured along the surface of the pitch cone.

Cone Gear (cone drive). Refers to a belt drive between two similar coned pulleys which by lateral movement provide a variable speed ratio.

Cone Plate (boring collar). It is a small bearing bolted to a lathe bed and carrying a circular plate or disc perforated with a series of tapered holes and fed forward in the spindle shaft, etc., when boring.

Cone Pully. A belt pulley which is stepped with two or more diameters to give different speed ratios having a similar pulley.

Conical Pivot. It is a pivot shaped as a cone which runs in a screw with a tapered hole, the taper angle being greater than the cone angle. The device has been used for the balance staff in alarm-clocks and in watches with pin-pallet escapement.

Connecting-rod. Refers to the rod connecting the piston or crosshead to the crank in a reciprocating engine or pump.

The 'forked connecting-rod' is having the crosshead enclosed in the forked end.

Connecting-rod Bolts (big-end bolts). Bolts which are securing the outer half of a split big-end bearing of a connecting-rod to the rod itself.

Constant-mesh Gearbox. It is a gearbox in which the pairs or wheels providing the various speed ratios have been always in mesh. The ratio may be determined by the particular wheel which gets coupled to the mainshaft by sliding dogs working on splines.

Constant-speed Propeller. Refers to a propeller the pitch of which gets controlled to vary automatically so as the maintain a constant rotational speed, which is desirable for the engine.

Constant Travel. Means the travel of a slide valve which cannot be varied for purposes of variable cut-off.

Constraint. The property which distinguished a mechanism from a kinematic chain.

Contact Breaker. A device which is used for repeatedly breaking and remaking an electrical circuit.

Contact Ratio. Refers to the ratio of the angle of rotation of a gear between the beginning and ending of contact of a tooth to the angle given by the fraction $360°/$(the number of the teeth in the gear).

Continuous Brake. Refers to a brake system in which operation at one point applies the brakes throughout a passenger train.

Continuous Mill. A type of rolling mill in which the stock passes through a series of pairs of rolls, undergoing successive reductions.

Continuous Path Motion. Motion of the end effector of a robot.

Contra-rotating Propellers. Two propellers which are mounted on concentric shafts with a common drive and rotating in opposite directions.

Contrate Gear (face gear, straight bevel rack). A gear having the tooth crests in one transverse plane and the roots in another transverse plane.

Contrate Wheel

(*a*) Refers to a toothed wheel with the teeth which is formed at right angles to the plane of the wheel for transmitting motion between two arbors at right angles.

(*b*) Refers to the fourth wheel in a watch with the verge escapement.

Control Jets. Jet of gas from pressure cylinders or some power source to control an attitude of a spacecraft in orbit.

Controllable-pitch Propeller. Refers to a variable-pitch propeller in which the blades could be set to predetermined pitch angles while rotating.

Conveyor. A device which is used for moving parts or materials from one place to another, usually over a short distance and/or from one level to another, generally on a continuous band called a 'conveyor belt'.

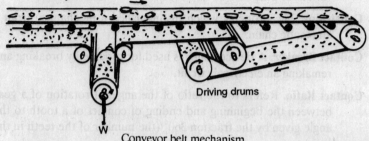

Conveyor belt mechanism.

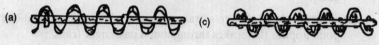

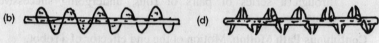

Conveyor Screws.

Conveyor Screws. Different types of screws which operate conveyors have been shown in Figure.

Coolant

   (*a*) A liquid (or gas) which is used as the cooling medium for an engine; for example, in the jackets of liquid-cooled piston-engines or as a film on the inner wall surfaces of combustion chambers and the nozzles or rocket-engines, or a molten metal, like sodium in hollow exhaust valves and certain atomic reactors.

   (*b*) Fluid which is used during machining to cool and lubricate the cutting or other tool and the work.

Cooling System. Refers to the system by which an engine or mechanism gets cooled by air or by a coolant.

    'Ducted cooling' has been obtained by constraining air to flow in ducts.

Cop. Refers to a conical ball of thread of yarn wound upon a spindle. It varies in size with the type and the count of yarn produced by the mule.

Copying Carriage. Refers to that part of a copying machine which travels along the model that is being copied.

Copying Machine. Refers to a class of machine for producing similar objects from a master pattern or template by using an engraving tool, end cutter or lathe tool, guided automatically by the pattern or template.

Corliss Valve. A steam-engine valve having an oscillation rotary motion over a port for admission of steam and its exhaust, the motion being controlled by an eccentric-driven wrist plate.

Cornish Engine. A massive type of beam engine which is used for pumping, originally single action, but later double acting and worked expansively.

Corrugated Expansion Joint. An expansion joint of the corrugated type has been shown in section in Figure 1. Corrugated diaphragms are used in pressure transducers. Figure 2.

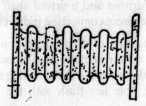 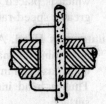

Fig. 1. Corrugated expansion joint (bellows type). **Fig. 2.** Cottered joint.

Cotter. Refers to a tapered wedge, rod or pin which is passing through a slotted hole in one member and bearing against the end of a second encircling member whose axial position is to be fixed or adjustable.

Cotter-pin (Split-pin). A split-pin inserted in a hole, as in a cotter, to disallow loosening under vibration.

Cotter Way. The slot cut in a rod to receive a cotter as shown by the outer shaft of Figure 2.

Coulomb (C). The unit of electric charge; it may be defined as the quantity of electricity transported in one second by a current of one ampere.

Counterboring. Means the boring of the end of a hole to a larger diameter.

Counterpart Racks. Racks, the teeth of which, when engaged, each exactly fill the spaces between the teeth of the other.

Counterpoise Bridge. A bridge in which the raising of a platform, roadway, etc., has been assisted by counterpoise weights.

Counterpoise Weight. A weight to give static balance (see balancing static), usually to ease the effort of moving the whole or part of a structure.

Counter (revolution counter). An instrument which is used for recording the number of operations performed by a machine or the number of rotations of a shaft.

Counter-reating Propellers. Refers to a pair of propellers in a ship which rotate in opposite directions but not on the same shaft.

Countershaft. Refers to an intermediate shaft in a line of shafting which is placed between a driving and a driven shaft to get a greater speed-ratio or where direct connection is difficult.

Countersunk Head. A head consisting of a truncated cone whose apex joins the main body of the screw, bolt, etc. The base of the cone has been in contact with a cone cut in the mating material. Thus a head inserted in a hole lies flush with the material surface.

Coupled Wheels. The wheels of a locomotive which get connected by coupling rods to distribute the driving effort.

Coupling

   (*a*)   A device which is used for connecting two vehicles.

   (*b*)   A device which is used for connecting railway coaches or trucks. It consists of two links united by a right- handed and left-handed screw.

Coupling Rod. A connecting-rod which is joining two cranks so that they work together as one.

Crab (crab winch)

   (*a*)   A jib-less hoisting crane having a snatch block or running pulley pendant from the barrel.

(b) The travelling lifting-gear of a gantry crane (see crane), mounted on a bogie and running on rails carried by the gantry.

(c) A claw coupling.

Crack Arresters (crack stoppers). Design features which, are incorporated into a structure to impede the propagation of a brittle crack. These have been particularly useful in large structures made by welding.

Crane. A machine which is used for hoisting and lowering heavy weights using gearwheels, chain barrel and chain.

Balance crane. It is a two-armed crane, one to take the load and the other counterpoise arrangements, which can be arranged as self-acting.

Cantilever crane. It is a crane in which the jib hangs out from the supporting member and is counterbalanced, such as is used for the transport of excavated materials from the bottom of a cutting to a spoil-bank, generally by skips.

Floating crane. It is a large crane carried on a pontoon as used in docks, etc.

Gantry crane. It is a travelling crane equipped with legs which support ir on rails at ground level.

Goliath crane. It is a giant travelling crane.

Hercules crane. It is a steam travelling crane with a horizontal swivelling jib used in harbour works for the setting of concrete blocks.

Horizontal crane. It is a portable steam balance crane with horizontal cylinders.

Hydraulic crane. It is a crane operated by hydraulic power

Jib crane. It is a crane with an inclined arm (a jib) attached to the foot of a rotatable vertical post, supported by a tie rod connecting the two upper ends, the chain running from a winch on the past and over a pulley at the end of the arm.

Derricking jib crane. It is a crane with variable radius of action obtained by changing the lengths of the tie rods between posts and jib.

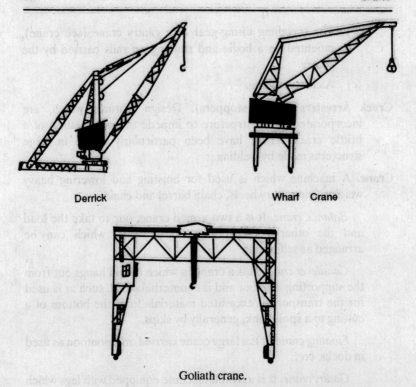

Derrick   Wharf Crane

Goliath crane.

Travelling jib crane. It is a jib crane having a travelling trolley on a horizontal jib which is mounted on the vertical post.

Lever-luffing crane. It is a jib crane in which during derricking or luffing the load can be moved radially in a horizontal path having consequent power saving.

Overhead travelling crane. It is a workshop crane which consists of a girder mounted on wheels running on rails fixed along the length of the shop near the roof, and traversing and lifting are usually done by power or may be done by hand. This crane is also termed as a 'shop traveller'.

Platform crane. It is a whip crane which is independent of any support at the top of the post.

Portal jib crane. It is a jib crane which is mounted on a fixed or movable structure with an opening directly under the crane to permit the passage of wagons, etc.

Post crane. It is a jib crane having the post supported on fixed pivots at the top and bottom. The radius of the circle in which the hook travels gets fixed when a tie-rod connects the post and boom; when ropes have been used the radius can be varried.

Rope crane. It is a travelling crane driven by an endless rope which travels at a very high speed so that minimum of power gets needed to lift a heavy load.

Steam crane. It is a crane which is operated by a steam engine.

Titan crane. It is a very large steam crane similar to a hercules crane but not provided with motions for slewing.

Tower crane. It is a rotatable cantilever crane which is pivoted at the top of a steelwork tower, either fixed or carried on rails. The lifting machinery and dead weights on the opposite side of the pivot balance the load.

Vertical crane. It is a steam crane with tall side frames.

Well crane. It is a fixed post crane with one-half of the post in a well with the lower end on a step and the fulcrum at ground level.

Wharf crane. It is a travelling or fixed crane on a quay which is having a fixed radius of action.

Whip crane (dutch wheel crane). It is a light derrick with tackle for hoisting but no gearing.

Crank. An arm which is attached to a shaft carrying at its outer end a pin parallel to the shaft.

Crank circle (crank path). The circle which is describing by the crankpin.

Crank Effort. The effective force which is acting on an engine's crankpin.

Crank Throw

   (*a*) Refers to the radial distance from the mainshaft to the crankpin and equal to half the stroke.

   (*b*) Refers to the webs and pin of a crank.

Crank Web. It is the arm of a crank which is usually of flat rectangular section.

Crankcase. Refers to the housing which encloses the crankshaft and connecting-rods.

Cranking. Hollowing of a tool immediately behind the cutting edge.

Crankpin. It is the pin which unites the web or arm of a crank with the connecting-rod of an engine or pump.

Crankshaft. It is main shaft of an engine, or other machine which is carrying a crank or cranks for the attachment of connecting-rods by the crankpins.

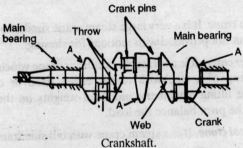

Crankshaft.

Creep

(a)   Refers to a slow plastic deformation of a metal under stress especially at high temperature.

(b)   Refers to the slow relative movement of two parts of a structure.

(c)   Refers to a slow relative motion of a belt over a pulley due to its varying extension and relaxation on either side of each pulley the belt encounters.

Creep Limit. Refers to limit of proportionality.

Creep Strength. Refers to the stress which will produce plastic deformation of a given metal at a specified rate of growth and at a given temperature.

Crescent (passing hollow). Refers to a circular notch in the periphery of the safety roller to allow the passing of the safety finger.

Crest. Refers to the surface of a screw thread which is connecting adjacent flanks at the top of the ridge.

Major (minor) crest truncation. Refers to the distance between the generators of the major (minor) cylinders or cones for the basic and design forms of the external (internal) screw thread.

Crimping Machine. A machine which is used for compressing a thin metal ring or cap into corrugations to reduce its diameter such as crimped caps used on bottle tops.

Crocodile Truck. A railway truck which is having a long open platform carried between a pair of four-wheeled bogie frames.

Cropping Machine. A shearing machine.

Cross-axle. A driving axle which is having cranks mutually at right angles.

Cross-cut Saw. A saw which has been designed for cutting timber across the grain with teeth shaped like an equilateral triangle.

Cross-section

   (*a*) Refers to the section of a body, typically an extrusion or I-beam at right angles to its length.

   (*b*) A drawing which is illustrating a cross-section.

Crossed Belt. It is a driving belt which passes from the upper side of one pulley to the lower side of another pulley, the pulleys revolving in opposite directions.

Crossed Rods. These are the eccentric rods of reversing engines which cross each other to join the ends of the slot-link, the centres of the sheaves being between the axle and the link. When they do not cross they are termed as 'open rods'.

Cross-arm Governor (parabolic governor). Refers to a governor in which the balls and the points of suspension of the rods have been on opposite sides of the central axis. The path of the ball gets arranged to the approximately that of a parabola.

Crosshead

   (*a*) Refers to a reciprocating member, generally sliding between guides, and making the connection from piston rod to connecting-rod in a reciprocating engine or pump.

   (*b*) Refers to the head of a screw with 'X' shaped recesses in it to take a matching shape of screw driver end.

Cross-slide. Refers to the slide or bridge which is carrying the tool on a planning machine or lathe which can get traversed at right angles to the bed of the machine and also raised or lowered.

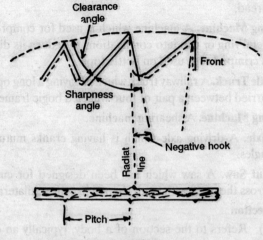

Cross-cut teeth.

Crown Gear. A bevel gear having a pitch angle of 90°.

Crown Wheel

    (*a*)   A bevel wheel having its teeth at right angles to its axis.

    (*b*)   The larger wheel of a bevel reduction gear.

    (*c*)   A wheel having gear-teeth in the ordinary spur gear position and ratchet teeth at right angles to the body of the wheel, as used in Swiss keyless mechanisms.

Crowned Pulley. A type of pulley whose circumferential surface has been convex to disallow lateral movement of the belt which drives it.

Crowing. Means the progressive reduction of tooth thickness towards the ends of a gear-tooth.

Crusher. A machine which is used for the mechanical subdivision of solids including a 'jaw crusher' which has a set of vertical jaws, one fixed and the other moved back and forth; A 'gyratory crusher' having inner and other vertical crushing cones, the inner with vertex upward and the outer with vertex downward; 'crushing rolls' consisting of two horizontal cylinders close together and rotating in opposite directions; hammer mills and ball mills.

Cumulative Pitch. Refers to the distance between corresponding point on any two thread forms of a screw thread which is measured parallel to the axis of the tread, whether in the same axil plane or not.

Cup Chuck (bell chuck). Refers to a hollow cylindrical chuck which is screwed to the nose of a mandrel in which small articles are held by screws of the walls of the chuck.

Cup Drum. It is a sheave-wheel whose rim gets recessed for individual chain links.

Cup Leather. It is a leather washer of U-shaped cross-section designed to prevent leakage of fluid past plungers and pistons in pneumatic and hydraulic machinery, the bicycle pump being an example.

Cup Wheel. Refers to a grinder in the form of a cylinder, the grinding being done on the revolving edge.

Cupping (tool). A punch which is used to form a hollow pressing from a blank.

Curb (curb ring). Refers to an internal ring of teeth on which a pinion, revolving in a fixed bearing on the upper portion of a crane, travels via suitable gearing and thus slews the crane.

Curb Pins (index pins). Refers to the pins which are fixed to the index to control the balance spring, making it vibrate faster or slower to regulate the watch.

Curling Tool. A tool which has been designed and made to curl the edge of a metal artical; it curls, coils, rolls, laps or bends the edge and the lap may get formed either internally or externally.

Cut. Refers to thickness of a metal shoving (chip) removed by a cutting tool.

Cut-off. Refers to the point at which a valve closes the port opening of an engine cylinder to the admission of steam. It is usually expressed as a percentage or fraction of the stroke, but sometimes as so many inches.

Cut-off Valve

(*a*) Refers to a separate slide valve which is fitted to control the admission of steam in a steam-engine when it is necessary or desirable to cut off the supply at a period earlier than half-stroke.

(*b*)   A valve actuated by the governor of certain gas engines to cut off the supply of mixture before the completion of the charging stroke, thereby regulating the quantity of mixture.

Cutter. A single-point tool which is used to a lathe, planner or shaper, or a multi-toothed cutter includes :

(*a*)   the 'profile type' which gets limited to straight-line cutting edges of pointed teeth, and

(*b*)   the 'form type' in which the face of each curved tooth gets sharpened.

>   *Rotary cutter.* A cutter which is rapidly rotated on its spindle.

>   *Side and face cutter.* A milling cutter, having teeth on the side as well as on the periphery, used for cutting slots.

Cutting Fluid. Any fluid which is used for lubricating a cutting tool and washing away chips and swarf.

Cutting Speed. Refers to the speed of the cutting tool relative to the workpiece generally defined as m/s or feet/minute.

Cutting Tools. Steel tools which are used for the machining of metals.

Cycle

(*a*)   Refers to the sequence of values of a periodic quantity throughout a complete period.

(*b*)   Refers to the sequence of operations in an internal-combustion engine namely, induction, compression, ignition and exhaust.

Cyclic Hardening. Hardening of material under repeated loading.

Cyclic Pitch Control. Refers to the control of a helicopter rotor by which the blade angle gets varied sinusoidally with the blade azimuth position; also termed as 'azimuth control'.

Cyclic Softening. Softening of some materials under repeated loading.

Cyclic Stress-train Curve. Refes to a curve of stress plotted against strain for repeated loading past the elastic limit of a material which makes a series of ever enlarging hysteresis loops.

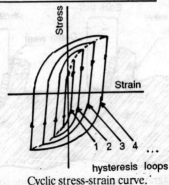

Cyclic stress-strain curve.

Cycloid. Refers to the curve traced on a plane by the motion of a point fixed on the circumference of a circle which rolls along a straight line.

Cycloidal Curves. Refers to the shapes of gear-teeth with profiles which have been cycloids, epicycloids or hypocycloids or combinations of the last two.

Cycloidal Gear. Gears having cycloidal teeth. When in mesh the faces of the teeth in one are epicycloids and the flanks of the teeth of the other have been hypocycloids, both curves being generated by same rolling circle.

Cycloidal Teeth. Refers to the teeth of gear-wheels whose flank profiles have been cycloidal curves.

Cylinder. The cylindrical (tubular) chamber in which the piston of an engine or pump reciprocates.

Cylinder Barrel. Refers to the wall of an engine cylinder.

Cylinder Bit (half round bit. D-bit). A boring tool having the section at the cutting face a semicircle. The cutting face has been sloped at an angle of about 4°.

Cylinder Block. Refers to the body of an internal-combustion engine in which the cylinders are located.

Cylinder Bore. Refers to the internal diameter of the cylinder of a piston-engine.

Cylinder Bore Gauge. Refers to a gauge, usually with a centralizing shoe to provide accurate location on each side of the gauging

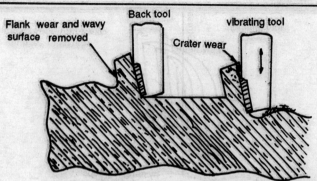

Diagrammatic action of two cutting tools.

point, and by this means, to give a true bore measurement, employing a transmission system to record on a circular dial.

Cylinder Cover. Refers to the end cover of the Cylinder of a compressor or of a reciprocating engine.

Cylinder Escape Valve. A spring-controlled valve which is fitted to the cylinders of marine engines.

Cylinder of Generation (of a gear). Refers to the pitch cylinder when meshed with its generating cutter.

Cylinder Head. Refers to the closed end of the cylinder of an internal-combustion engine. It may be integral with the cylinder but more generally it is removable and contains the valves.

Cylinder Reference. In helical and spur gears, the right circular cylinder on which the normal pressure angle is having a specified standard value.

Cylindrical Gauge. Refers to a plug gauge of a defined length which is suitable for veryfying both depth and diameter of hole.

Cylindrical Grinding. Using a high-speed abrasive wheel to finish accurately cylindrical work by rotating the work in the head-stock and automatically traversing the wheel along it under a copious flow of coolant.

δ. Greek lower case delta. Also, refers to the symbol for deflection (amongst others).

Db. Decibel.

DB Pull. Draw Bar pull.

DHN. Abbreviation of Dynamic Hardness Number.

DPN. Abbreviation of Diamond Pyramid Hardness Number.

D Slide Valve. A slide valve of D section which slides on a flat face in which ports are cut.

DTI. Abbreviation of Dial Test Indicator.

Damper. A device which is used for dissipating energy in a mechanical system by the suppression of vibrations of unfavourable non-linear characteristics.

    Blade damper. A damper which is used to prevent the hunting of a helicopter rotor.

    Roll and yaw dampers. Dampers which are used for the suppression of rolling and azimuth oscillations of an aeroplane respectively.

    Shimmy damper. A damper which is used for the suppression of shimmy.

    Friction damper. A mass which is frictionally driven from a crankshaft at a point remote from a node to dissipate the energy of vibration in heat.

    Vibration dampers. Dampers fitted to an engine crankshaft to suppress or reduce stresses resulting from torsional vibration at critical running speeds.

    Torsional vibration damper. It is a flywheel mounted on a shaft with the relative motion damped by viscous friction.

Tuned torsional vibration damper. It is a flywheel coupled to a shaft by a spring to from a resonant system effective at frequencies near its natural frequency.

Damping. Refers to the process by which the energy of a vibrating system gets dissipated.

Damping coefficient. Refers to the constant coefficient of the velocity term, x, in a motion defined by the differential equation $mx + cx + kx = 0$, where m is the mass and k is the stiffness of the system.

Damping ratio. Refers to the ratio of the damping coefficient to the critical damping coefficient.

Critical damping coefficient. Refers to the smallest value of the damping coefficient which is required to prevent vibration.

Coulom damping. Damping in which the force opposing a motion is having a constant magnitude.

Internal damping. Damping intrinsic to the materials.

Magnetic damping. Damping due to eddy currents which are set up by the movement of a system in a magnetic field.

Structural damping. Damping due to the total effect of a built-up structure.

Viscous damping. Damping in which the opposing force has been proportional to the velocity.

Dashpot

(a)   Refers to a damping device which consists of a piston and cylinder whose relative motions gets opposed by the fluid friction of a liquid or of air.

(b)   A cylinder used in steam-engines fitted with trip gears for closing the admission valves suddenly as soon as they get released by the trip.

Datum

(a)   A point from which all measurements could be made.

(b)   A line from which all measurements could be made.

(c)   A horizontal plane from which all vertical measurements could be made.

Datum Level. Refers to a base line of a section from which all heights and depths could be measured.

Datum Line. Refers to a defined line or base from which dimensions have been taken or calculations are made. It establishes an exact geometrical reference.

Daylight. In a machine press, the distance between bed and the lowest position of the face of the ram.

De Dion Axle. It is a motor vehicle rear suspension halfway between a normal back axle and an independent suspension. The final differential drive gets bolted to the frame of the vehicle. The stub-axles get driven through universal joints adjacent to the final differential drive and to the wheels which are supported by leaf springs from the vehicle frame.

De Laval Turbine. Refers to an early single-wheel impulse turbine.

De-cluch. Refers to disengagement of a clutch.

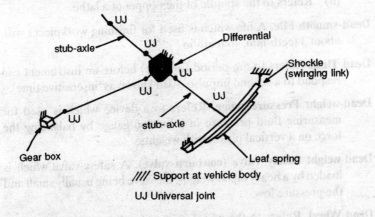

De dion Axle.

Dead Angle. Refers to the angle of movement of the crank of a steam-engine during which the engine will not start when the stop valve gets opened due to the ports being closed by the slide valve.

Dead Axle. Refers to an axle which does not rotate with the wheels carried by it.

Dead-beat Escapement. An escapement having no recoil of the escape wheel. It is used for regulators.

Dead-centre Lathe

(a) Refers to a lathe in which the work alone revolves between dead-centres: the mandrel does not revolve.

(b) Refers to a small instrument-maker's lathe in which both centres gets fixed, the work alone being revolved, *e.g.*, by a small pulley mounted thereon.

Dead Centre (dead points). Refers to the least and greatest extension of a piston or a crank, where it exerts no effective power, the piston rod being coaxial with the cylinder.

Dead End (poppet). The tail stock of a lathe having a dead spindle and a back centre.

Dead Spindle

(a) Refers to the arbor of a machine tool that does not revolve.

(b) Refers to the spindle of the poppet of a lathe.

Dead-smooth File. A life which is used for finishing workpieces with about 3 teeth/mm, 76 teeth/in.

Dead Time. Refers to the period (or time) before an instrument can respond to a second impulse. Also known as 'insensitive time'.

Dead-weight Pressure-gauge. Refers to a device which is used for measuring fluid pressure in a burdon gauge by balancing the force on a vertical piston with weights.

Dead-weight Safety Valve (cowburn valve). A safety-valve which is loaded by a heavy metal weight, the valve being usually small and the pressure low.

Dead Wheel. Refers to the wheel in an epicyclic gear around whose centre the remainder of the gear (or train) is revolving.

Decibel. It is the unit of difference of power (or noise intensity) level which is measured logarithmically. If P_1 and P_2 are two amounts of power and N the number of bels, then $N = \log_{10} (P_1/P_2)$. A decibel has been equal the one tenth power of a bel.

Dedendum. Refers to the whole depth of a wheel tooth less the appropriate addendum.

Dendum angle. Refers to the difference between the pitch angle and the root angle of a bevel gear.

Dedendum (screw thread). Refers to the radial distance between the pitch and minor cylinders (or cones) of an external thread; the radial distance between the major and pitch cylinders (or cones) of an internal thread.

Degress of Freedom. May be defined as the minimum number of co-ordinates which are need to specify the possible motion of a mechanical (or other) system.

Delay Period

(*a*)   Refers to the time or crank angle between the passage of the spark and the pressure rise in a piston engine.

(*b*)   Refers to the time between fuel injection and pressure rise in an oil engine.

Demagnetizer. A device which is used for removing residual magnetism from metalparts of a mechanism or from a piece of work held by a magnetic chuck or from some other cause.

Depth Gauge. A gauge or tool which is used for measuring the depths of holes by the use of a narrow rule of cylindrical rod working within, and at right angles to, a crossbar.

Derailer. Refers to an arrangement of rails for turning-off deliberately runway trucks from the tracks.

Derailleur Gear. Refers to a method of changing the gear ratio of chain-driven sprocket wheels. This is realized by guiding the chain from a sprocket wheel of one size to one of different size, while maintaining some tension with a tensioner and permitting uninterrupted power transmission. It has been used in racing bicycles where there may be two sprocket wheels at the pedals and five at the rear wheel giving, with two independent chain guides, a total of ten gear ratios.

Derrick. Refers to a type of crane in which the radius of the jib can get altered by means of ropes or chains passing over the top of the post or mast. Also termed as a 'Derricking jib crane'.

Derriking. Refers to the raising or lowering of the jib of a derricking jib crane.

Desaxe Engine. The term used for a reciprocating engine machanism in which the line of stroke does not pass through the axis of the crankshaft. The obliquity of the connecting-rod gets diminished during the forward stroke, the pressure on the guide bars gets reduced and the turning moment is slightly more uniform; the return stroke gives less advantage; on the whole there occurs a slight gain.

Desmodromic. Refers to a method of operating poppet valves so that they are positively opened and positively closed by the mechanical linkage transmitting the timing. The return does not depend on a spring, although one may provide assistance. These mechanisms are usually cam-driven and are designed for maintaining full valve control during high-speed operation. Their chief use has been in motorcycle engines.

Detent Cams. Small roller cams fitted to the follower of a larger cam having a detent notch.

Detent Notch. An axial groove in the peripheral face of a cam.

Detonation. Refers to the spontaneous combustion of part of the compressed charge within the cylinder of an internal combustion engine and its attendant sharp noise or knock.

Detruding. Refers to thrusting or forcing down or pushing down forcibly, as when a hole gets punched in a plate, leaving the plate strained.

Deruslon. Shear Strain.

Detuner (dynamic damper). Refers to an auxiliary vibrating or rotating mass which is driven through springs to modify the vibration characteristics of the main system to which it gets' attached, thus eliminating a critical vibration or critical speed. An auxiliary mass controlled by a non-linear spring has been one type of detuner.

Diagonal Winch. A steam winch the cylinders of which are kept diagonally on the side frames.

Diagram Factor. Refers to the ratio of the mean effective pressure developed in a steam-engine cylinder to its theoretical value; in practice it is varying between 0.5 and 0.9.

Dial. Refers to the observable part of an indicating instrument with a moving pointer indicating a reading on a graduated scale.

Dial Gauge (dial test indicator). It is a sensitive measuring instrument, which indicates small displacements of a plunger in thousandths of an inch by a pointer moving over a circular scale.

Dial Test Indicator (see dial gauge). Also called colloquially 'clock'.

Diametral Pitch (manchester principle). The number π which is divided by the spacing of adjacent teeth in inches.

Diametral normal pitch. The number π which is divided by the normal pitches in inches.

Transverse diametral pitch. The number π which is divided by the transverse pitch of a gear in inches; also the number of the teeth in the gear divided by the diameter of the reference circle in inches. This is sometimes termed as the 'diametral pitch'.

Diamond-tip Turning Tool. A typical turning tool having a cutting diamond tip.

Diamond Tool (diamond point). A name assigned to a tool when the surface of the cutting plane gets formed like a lozenge for diamond.

Diamond Wheel. An abrasive wheel having diamond particles bonded in it.

Diaphragm Pump. A pump in which a flexible diaphragm is having a reciprocating rod of short stroke attached at its centre.

Diaphragm Valve. Refers to a valve which relies upon the deflection of a flexible diaphragm, by fluid pressure applied, to shut off the fluid flow

Die

(a) Refers to an internally threaded steel block having cutting edges for producing screw threads. The larger dies are made in halves and set in a die stock. The block is termed as a 'screwing die' or 'chaser'. The diameter of the hole can be reduced by means of a pair of screws or increased by a third, central, pointed screw. For manual thread cutting, handless get fitted to the die holder and the correct depth of thread would be obtained by progressively closing the die.

(b) A metal block having a correct internal cross-section through which hot metal gets rammed when making

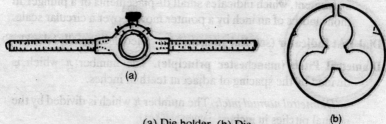

(a) Die holder, (b) Die.

extrusions. Cold metal and warm themoplastics are also extruded.

(c) When piercing is followed by blanking the die is termed as a 'progressive die'.

(d) A male member of the correct shape for stretch forming.

Die Chuck. A small two-or-three-jaw independent chuck.

Die Engraving. Engraving of dies used in stamping and press operations such as coin making and printing high-class stationery.

Diehead (die head; die box). Dieheads have been commonly used on the turret of a capstan lathe where they automatically form an

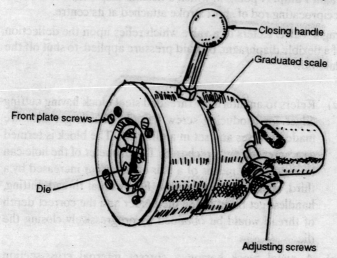

Self-openinng diehead.

external thread on the workpiece up to a pre-set length when the diehead opens and enables withdrawal of the tool from the work.

Diesel Cycle. Refers to the cycle of a compression-ignition engine in which air is allowed to be compressed, heat added to constant pressure by injecting fuel into the compressed charge, expanded to do work on the piston and the products exhausted, either in a four-stroke or a two-stroke cycle. It is involving reversible heating and reversible cooling at constant pressure.

Diesel Engine. It is a compression-ignition engine in which the oil fuel gets injected into the heated compressed-air charge, originally by a blast of air. It is a two-stroke or four-stroke engine in which the fuel is ignited by the heat of compression.

In the 'hot-bult type', ignition is carried out by means of a very hot bulb-shaped surface, and the operations of compression and fuel injection are carried out separately.

Diesel-electric Locomotive. A locomotive in which a diesel engine is able to drive an electric generator and the latter supplies current to electric motors which are connected to the driving axles.

Differential. Creating a difference.

Differential cam. Refers to an arrangement of cams having different outlines in the valve gearing of gas-engines, the cams sliding upon a shaft under the control of a governor to come into contact, in turn, with the roller and hence to very the admission to suit the load.

Differential gear (differential). Refers to an assembly of bevel or spur gear wheels having two co-axial shafts and a third co-axial member with a rotation proportional to the sum or difference of the amounts of rotation of the other two.

Differential lever (flooting lever). A rigid link which is carrying three pivots, the third of which is having a displacement dependent on the input displacements of the other two.

Rack-type differential. A linear diaplacement mechanism which has two racks engaging with opposite sides of a pinion wheel so that the linear movement of the wheel axis has been half the algebraic sum of the two linear displacements of the racks.

Synchro-control differential transmitter. A synchro having mechanically positioned rotor which is used for transmitting information corresponding to the sum of the difference of the synchro and electrical angles.

Synchro-torque differential receiver. A synchro-having freely turning rotor, which is able to develop a torque dependent on the difference between the two electrical angles received from its connected torque transmitters.

Synchro-torque transmitter. A synchro having a mechanically positioned rotor which is used for transmitting electrical information corresponding to the rotor position

Differential Motion. Refers to the motion of one part to and from another in a mechanical movement, the speed (velocity) of the driven part is equal to the difference of the speeds of the two parts connected to it.

Differential Pulley Block (chain block). An endless chain is running over two sheaves and a movable load pulley below. The sheaves get fixed together on one axis and the mechanical gain would be proportional to the difference in diameter of the two sheaves. Rotation of the sheave axis by a hanging loop shortens a second loop supporting the load pulley to give a large advantage.

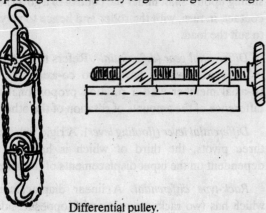

Differential pulley.

Differential Screw (compound screw). A screwed spindle which works within a nut and is also threaded externally for the reception of another fixed screw of the same hand but of slightly finer pitch. With the nut attached to the diehead of a press this device provides a high pressure through the prolonged action of a small force.

Differential Windlass. A windlass in which the power exerted has been due to the difference between the velocity of the rope upon two drums of unequal diameters.

Diffuser. The term used for chamber in which kinetic energy is converted into pressure energy by a gradual increase in the cross-sectional area of the flow of a gas.

Digger. A mechanical excavator having a digging shovel.

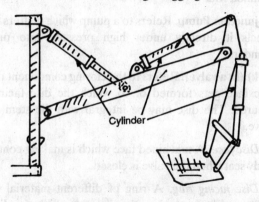

Mechanical digger.

Dimension. Refers to an element whose size gets specified in a design such as a length or an angle, and the word 'element' which it also refers to weights, capacities, areas etc.

Auxiliary of reference dimension. A dimension which is given solely for information or convenience of reference.

Constructional dimension. A dimension which defines a positional or angular relationship between two or more features, or the form of a surface including a profile.

Datum dimension. A dimension which is able to fix the position of a datum plan, line or point.

Dip Stick. A rod which is inserted into a container, frequently an engine sump, and marked to show the quantity of fluid present.

Dipper (or dipper-bucket) Dredger. A dredger having a single large bucket at the end of a long arm which can get swung in a vertical plane by gearing.

Dipping. Immersion in a path of liquid to produce a surface treatment such as pickling or galvanizing.

Direct-acting Engine. Refers to a steam-engine in which the action of the piston gets transmitted directly to the crankshaft.

Direct-acting Pump. Refers to a steam-driven reciprocating pump with the steam and water pistons carried on opposite ends of a common rod.

Direct-injunction Pump. Refers to a pump which matters the fuel and injects it directly under high pressure into piston-engine cylinders.

Disc (disk) (in a valve). Refers to the closing component on which the disc face gets formed or to which the disc facing ring gets secured. The disc may be integral with the stem of a needle valve.

Disc face. A machined face which is making contact with the body seat when the valve is closed.

Disc facing ring. A ring of different material to the disc, permanently secured to the disc, on which the disc face gets machined.

Disc guide. Refers to that part of a valve in which the disc or disc holder gets guided.

Disc guide pin. Refers to the pin which engages with the disc guide.

Disc guide wings. Refers to that part of the disc assembly in the form of wings which guides the disc to the body seat.

Disc Brake. Refers to a brake in which the friction is got from pads acting upon a disc on a vehicle's wheel or on the landing wheel of an aircraft's undercarriage or, similarly, for braking machinery.

Disc Crank (crank plate, balanced crank). Refers to a crank of circular outline on which the metal has been sometimes so disposed as to balance the varying motion of the connecting-rod.

Disc Loading

    (*a*) Refers to the thrust of a propeller divided by the disc area.

    (*b*) Refers to the lift of a helicopter-rotor divided by the disc-area.

Disc Separator. It is a sorting machine for grain or seeds of various sizes which pass between revolving iron discs on horizontal shaft. The faces of the discs as having numerous small indents which remove the unwanted seed by discharging it at a higher level.

Disc Valve. It is a light steel or fabric disc resting on a ported flat seating. It is used as a suction and delivery valve in pumps and compressors. Steel discs are usually spring-loaded.

Disc Wheel. A wheel having a solid disc of metal connecting the hub and rim.

Disc-and-drum Turbine (combination turbine, impulse reaction turbine). Refers to a type of steam-turbine having a high-pressure impulse wheel, followed by intermediate and low-pressure reaction blading, mounted on a drum shaped rotor.

Discharge Valve

    (*a*) A self-acting valve which is used for controlling the rate of discharge of a fluid from a pipe or centrifugal pump.

    (*b*) Any delivery valve.

Disengaging Clutch (coupling). A clutch which is used for throwing out of gear a line of shafting or a train of wheels.

Disengaging Gear. Refers to the mechanism for operating a disengaging clutch.

Disintegrator (disintegrating mill). A mill which consists of fixed and rotating bars in close proximity for reducing lump material to granular product.

Displacer Piston. Refers to auxiliary piston in some gas-engines which is used for expelling the residual products of combustion from the cylinder.

Distributor Rollers. Refer to the rollers which distribute ink on the inking table in a cylinder printing press or those carrying ink in a duplicating machine.

Dither Mechanism. Refers to a mechanism which is used to remove stiction by providing an oscillatory motion of small amplitude between two relatively moving parts.

Divergent Nozzle. Refers to a nozzle with a cross-section increasing continuously from entry to exit. It is used in compound impulse turbines.

Divided (axial) Pitch. Refers to the axial distance between corresponding points (such as centres of successive helices) on successive threads of a multiple-threaded screw.

Dividers. These are a pair of hinged pointed arms which, like a compass, can be hand adjusted for measuring, transferring measurements, or scribing circular arcs on work.

Dividing Engine. An instrument which is used for making or engraving accurate subdivisions on seales; it may be linear, circular or cylindrical. The linear machine is having a carriage, adjusted by a micrometer screw and holding a making tool, which gets moved on a very accurate lead screw. It is sometimes called a 'dividing machine'.

Dividing Head (indexing head). The term used for an attachment which is used on a milling-machine table for accurately dividing the circumferences of components for grooving or fluting, gear-cutting, the cutting of spliners, etc. Other machine tools are having similar attachments.

Double Vane. It is a double cup having central cutwater which splits the jet impringing on the vanes of a Pelton wheel and turns the water through an angle of about 160° to clear the wheel exist.

Doffer. It is a wire-covered cylinder of a carding engine which is able to remove fibres from the wire-covered surface of the main cylinder.

Doffing Comb. Refers to a steel blade which is extending across and oscillating cover, the doffer of a carding engine to strip the carded material in the form of a sheet or web.

Dog

    (*a*) Refers to a jaw of a chuck.

    (*b*) Refers to the carrier of a lathe.

    (*c*) Refers to a pawl (d).

    (*d*) Refers to an adjustable stop in a machine tool.

Dog-chuck (jaw chuck). A lathe chuck which is usually with four independent jaws or dogs.

Dog Wheel. See ratchet wheel.

Dolly Bolt. Refers to a bolt which is passing through the laminated springs on a vehicle and continued into the axle to prevent the U-bolts from slipping.

Dome. A domed cylinder which is often attached to a locomotive boiler and acts as a steam space and to house the regulator valve.

Donkey Boiler. It is a small vertical boiler for suppling steam to drive small engines and machinery, especially on board ship.

Donkey Engine. An auxiliary engine which is for pumping and light work, especially on board ship.

Donkey Pump. A small steam reciprocating pump which is used on board ship.

Dorr Mill. A tube mill which is designed for operation as a closed-circuit wet-grinding unit.

Double-acting Engine. The term used for a reciprocating engine in which the working fluid acts alternately on each side of the piston. All steam-engines were originally single-acting, but all have been now double-acting. Very few internal-combustion engines have been double-acting.

Double-acting Piston. A piston which is acted upon on both sides alternately by gaseous or fluid pressure.

Double-acting Pump (double-action pump). Refers to a reciprocating pump in which both sides of the piston are acting alternately to give two delivery strokes per cycle.

Double-acting Steam Hammer. It is a steam hammer which allows steam above and below the piston.

Double-angle Drill Point. A drill point having the clearance face ground in the form of two cones. The double-cone is able to decrease the side clearance angle along the outer cutting edge, l_0, and increases the value of the tip angle compared with a conventional single-angle drill point. A longer tool life, quietness, small-axial force for given diameter, d, and better machining accuracy for a given power input, has been obtained.

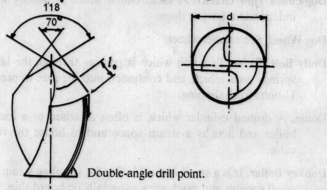

Double-angle drill point.

Double-beat Valve (cornish valve). Refers to a hollow cylindrical lift valve having two seating faces at the two ends of only slightly different areas. When exposed to pressure the valve gets nearly balanced and easily operated. It is used for controlling high-pressure fluids.

Double Cards. Indicator cards which are taken from both ends of the cylinder of a steam-engine.

Double-cutting Drill. Refers to a drill which is ground to cut with equal facility in a right-or left-hand direction. The cutting edge is lying in the longitudinal axis and the angles are symmetrical on the two sides.

Double-cylinder Engine. Refers to a steam-engine having two cylinders, one piston being at half and the other at full stroke when the cranks have been at right angles to each other.

Double Disc (gate) Valve. Refers to a gate valve in which the gate is consisting of two discs forced apart by a spreading mechanism at the point of closure against both parallel body seats to ensure effective sealing of the valve without the help of the fluid pressure.

Discs. Refers to the components, attached to the stem of the spreading mechanism, on which the disc faces get machined.

Double Driver Plate. A driver plate having two pins which engage and drive a carrier attached to the revolving work. It has been used for work requiring special accuracy.

Double-ended Bolt (weish bolt). A bolt which is screwed at each end for the reception of a nut having a central portion with flats or a tommy-bar hole.

Double-ended Machine. Refers to a punching and shearing machine, the two sets of operations being carried on at the same time.

Double-entry Compressor. A centrifugal compressor having two-vaned impellers mounted back-to-back so that air get admitted from ends of the rotating member and ejected in the central plane.

Double-flow Turbine. Refers to a turbine in which the working fluid is entering at the middle of the casting and flows axially towards each end.

Double-helical (Herringbone) Gear. A gearwheel which is having teeth of helical pattern in opposite directions like a 'herringbone', thereby eliminating axial thrust.

Double-ported Slide Valve. Refers to a slide valve in which steam is admitted through two steam ports at each end of the cylinder face, thereby reducing the travel of the valve.

Double Purchase. A lifting arrangement which consists of two pinions and two wheels.

Double-roller Safety Action. A lever escapement having two rollers, one carrying the impulse pin and the other being used for the guard finger.

Double-row Radial Engine. A radial engine having two rows of cylinders which are arranged radially one behind the other and operating on two crank pins $180°$ apart.

Double Shaper. Refers to a shaping machine for a double set of operations having two rams and two tool boxes; *e.g.*, one operation being for straight and the other for circular cutting.

Double Sleeve Valve. Refers to a sleeve valve arrangement as shown in Figure which illustrates as open inlet port.

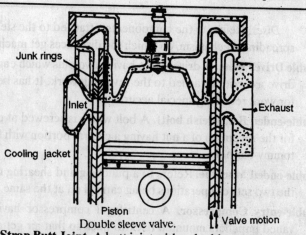

Junk rings

Inlet

Exhaust

Cooling jacket

Piston

Double sleeve valve.

Valve motion

Double Strap Butt Joint. A butt joint with a double strap, one on each side, riveted, bolted or welded together.

Double-threaded Screw (two-starts thread). A screw having two threads, half the true pitch apart. It gives an increase in the rate of travel.

Double Thrust Bearing. A thrust bearing which is used for taking axial thrust in either direction.

Dovetail Cutter. A specially shaped rotatory cutting tool which is used for shaping dovetail grooves. The grooves have tapered sides so that the mating part or taper headed bolt locates firmly when drawn tightly together inside the groove.

Dowzieme. It is a unit of length for the height or thickness of a watch movement. It is equal to a twelfth of a ligne, equal to 0.188 mm.

Drag Conveyor. A conveyor which is used for feeding loose material along a trough by means of an endless chain having wide links carrying projections on the wings.

Drag Link

(*a*) Refers to the suspension rods or links of a valve gear which are controlling the forward and backward gear of an engine.

(*b*) Refers to a link for connecting and disconnecting the cranks of coupled engines.

(*c*) Refers to a rod by which the link motion of a steam engine which are moved for varying the cut-off.

(*d*)   Refers to a link from the drop arm of the steering-gear of a motor-vehicle which acts on the steering-arm through ball joints at its ends.

Drag Surface. The forward face of a screw propeller which is used for ship propulsion.

Draw-bar (drag-bar)

(*a*)   The bar which is connecting a locomotive to the rolling stock and by which the tractive force is transmitted.

(*b*)   The bar which is connecting a tractor to a trailer in road vehicles; it is usually part of the trailer.

Draw-bar Cradle. A closed frame or link which is used for coupling the ends of draw-bars.

Draw-bar Plate. A heavy transverse plate through which the draw- bar is attached to a locomotive.

Draw-bar Pull

(*a*)   Refers to the tractive effort of a locomotive drawing a train in specified conditions.

(*b*)   Refers to the pull exerted by a tractor on its trailer.

Draw-bar Spring. A shock-absorbing spring which is fitted between the draw-bar and the frame of a railway carriage, or the springs fitted within the draw-bar of a trailer, sudden changes in draw-bar pull.

Drag Plate. The plate on which dies are supported in drawing operations.

Drawability. Drawability of a material for extrusion may be expressed by dividing the blank diameter minus the die diameter by the blank diameter.

Drawing

(*a*)   Refers to the representation by lines on paper of machines, etc., to get manufactured.

Assembly (sub-assembly) draing. A drawing which is limited to an individual assembly.

Detail drawing. A drawing which gives manufacturing requirements for a specific detail or details.

General arrangement drawing. A drawing of a complete finished product which shows (*a*) the components which make up the final assembly, and (*b*) the means of identification with other drawings.

Operation drawing. A document which gives the procedure to get adopted by the operator in carrying out the operations to make a specified part.

(*b*)   Refers to the manufacture of wire, tubing, etc., by pulling or drawing the material through dies of progressively decreasing size.

Dredger. Refers to a chain of buckets, scoops or suction pumps which is used for removing sand, alluvial deposits, etc., from under water and loading the material on to the barge or raft carrying the necessary machinery or on to another vessel.

Dredger Excavator. Refers to an excavator which is similar to the bucket dredger but designed to work on land.

Dresser. A tool which is used for facing and grooving millstones for truing grinding wheels.

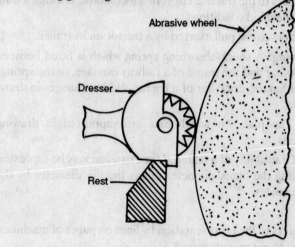

Dresser

Drift

(*a*)   A metallic wedge which is used to free two mating parts; such as two morse tapers, where the drift passes through slots in the outer one and engages the tang of the inner one displacing it axially.

(b) Drift or elastic drift is the term which is used on occasions to describe creep at normal temperatures when under constant elastic stress.

Drill

(a) A tool which is used for boring cylindrical holes with cutting edges at one end and having flats or flutes for the release of chips.

(b) Refers to a compressed-air-operated rock drill.

(c) A straight steel bar having a shank at one end a cutting edge at the other.

(d) The machine which is used to drive a drilling tool.

Drill Bush. A bush of hardened steel which is used to guide drills and reamers, enabling accurate positioning of holes in the workpiece when the bushes get held in a jig or fixture.

Drill Chuck. A self-centring chuck, usually having three jaws. It is used for holding small drills by contraction of the chuck with an internally coned sleeve encasing the jaws.

Drill Feed. A mechanism which is used for feeding a drill into the work.

Drill Plate (boring flange). Refers to a circular plate which is fitted over the nose of the lathe mendrel to take the pressure of a workpiece which is being drilled.

Drill Socket. A tapered sleeve which is bored out to a standard size to receive the shank of a twist drill.

Drilling Machine. A machine tool which is used for drilling holes. It is usually composed of a vertical standard with a table for supporting the work and an arm provided with bearings for the drilling spindle. Several drilling spindles have been frequently operated together on the same machine.

Masonary drill. Refers to a twist drill with hardened inserts forming the cutting edges.

Pillar drill. Refers to a drilling machine with brackets on a pillar to carry the spindle and table, the latter sliding on the pillar.

Radial drill. Refers to a large machine with the drilling head movable along a rigid horizontal arm, carried by a pillar.

Sensitive drill. Refers to a small machine with the drill fed into the work by a hand lever directly attached to the drilling spindle, thereby maintaining a sensitive control of the rate of drilling.

Driven Gear. In any pair of meshed gears, this gear refers to the one which is receiving movement, and therefore power, from its partner.

Driver Chuck. Refers to a point chuck in a mandrel for carrying work which is being turned between centres.

Driver Plate. A disc which is screwed to the mandrel nose of a lathe and carrying a pin which engages with and drives a lathe carrier attached to the work.

Driving Drum. Refers to a power-driven drum which transmits its energy to other parts of a mechanism, such as a conveyor belt or a bobbin.

Driving Gear

(a) Refers to any system of gears, belts, pulleys, shafting, etc., through which power gets transmitted to a machine.

(b) In any pair of meshed gears, this refers to the gear which is passing movement to its partner.

Driving Springs. The springs which are carrying the axle boxes of a locomotive's driving axles.

Driving Wheel

(a) Refers to the first member of a train of a gears.

(b) Refers to a wheel on a driving axle of a locomotive or road vehicle.

Drop Arm. Refers to an arm actuated by the steering-gear which converts rotary motion of the steering shaft to turning of the wheels.

Drop Test

(a) Refers to an environmental test in which the retardation gets varied to give the desired *g* force on the equipment under test in a drop test-rig. Retardation could be controlled pneumatically, hydraulically or by use of a suitable diameter of spike which enters a pad of lead.

(b) Refers to the dropping of a steel tyre on a steel rail from a specified height, varying with tyre diameter, as a strength test.

Drop Valve. A conical-seated valve with rapid operation by a trip-gear and return spring.

Drop Worm. A worm which can be dropped out of engagement with its wheel.

Dropping Valve. A valve which is used for reducing the supply pressure by a constant amount.

Drum

    (*a*)   Refers to the rotor of a reaction turbine.

    (*b*)   Refers to a cylinder on which lace threads get wound to a definite length, for transfer to brass bobbins.

Drum Scanner. A rotating drum which is used in mechanical scaning systems in television and carrying the needed picture-scanning elements.

Drunken Thread. Refers to a screw thread in which the helix angle varies along its length.

Dry-blast Cleaning. Refers to the blasting the surface of a workpiece with abrasive material travelling at a high velocity, including sand-blasting.

Dual Ignition. Refers to the use of two sparking plugs per cylinder eacy supplied from a separate electrical source.

Duct. A pipe or channel which is used for supplying a fluid, *e.g.*, for lubrication.

Ducted Fan. Refers to a fan which functions inside a duct. In a turbojet engine, the blades of the turbine get lengthened so that they extend into an annular space outside the engine, the fan being driven by the low-pressure turbine, independent of the high-pressure turbine of the main engine.

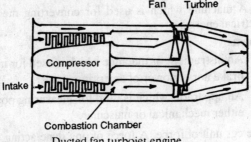

Ducted fan turbojet engine.

Ducted Propulsor. A propeller which is used for propelling under-water vehicles.

Dukey

(*a*)   Refers to a train of tubs which are travelling on an inclined haulage road underground.

(*b*)   Refers to a platform on wheels for tube or trams to get lowered on steep self-acting inclines.

Dummy Piston (balance piston). A disc kept on the shaft of a reaction turbine so that steam pressure could be applied on one side to balance the end thrust.

Duplex Carburetter. Refers to a carburetter in which two mixing chambers have been fed from a single float chamber.

Duplex Lathe. A lathe which uses two cutting tools simultaneously, one on each side of the work.

Duplex Planing Machine. A planing machine having two beds and two tables.

Duplex Pump. A pump having two working cylinders side by side.

Dwell. Refers to the angular period of a cam during which the cam follower gets allowed to remain at a given lift.

Dynamic Balance. Refers to the condition wherein centrifugal forces due to rotation produce neither couple not resultant force in the shaft of the rotating part, such as a flywheel, rotor or propeller.

Dynamic Hardness Number (rebound hardness). The number which is given by a Herbert pendulum or a Shore Scleroscope.

Dynamic Load (live load). A rolling or a moving load.

Dynamic Model. A free-flight aircraft model, a wind-tunnel model or a model of a seaplane hull (or float) with dimensions, masses and moments of inertia correctly proportioned, so that the model will simulate correctly the full-scale behaviour.

Dynamo. A machine which is used for converting mechanical into electrical energy.

Dynamometer

(*a*)   An instrument or machine which is used for measuring the brake horse-power of any prime mover.

(*b*)   An apparatus which is used for measuring power or force, either mechanical or mascular.

Dyne. The cgs unit of force. A force of one dyne acting on a mass of one gram imparts to it an acceleration of one centimetre per second per second. (1 dyne = 10^{-5} newtons).

$$\boxed{E}$$

E. Symbol for young's modulus. Also the symbol for energy.

    (a) Symbol for coefficient of viscosity, $\eta = \mu/\rho$.

    (b) Symbol for efficiency.

EAROM. Electrically Alterable Read Only Memory (used in robots).

EPROM. Electrically Programmable Read Only Memory (used in robots).

Eff. Symbol for efficiency.

EHp. Symbol for Effective Horse-power.

Eng. Engine.

Ear (lug). Refers to a permanent projection on an object for its support or for the attachment of another to it by a pivot.

Early Cut-off

    (a) Any cut-off which is shorter than one half the stroke in a steam-engine cylinder.

    (b) Refers to the occasion when the motors of a rocket or a launcher vehicle switch themselves off before the programmed time.

Eccentric

    (a) Not concentric.

    (b) A mechanism which is used for converting the rotary motion of a crankshaft into a reciprocating rectilinear motion, used chiefly for short throws.

    (c) Refers to a crank, in which the pin diameter gets exceeded the stroke resulting in a disc eccentric to the shaft, such as is used for operating steam-engine valves, pump plungers etc.

Eccentric Crank. A crank in a locomotive valve gear, which gets substituted for the eccentric.

Eccentric Key. A key recessed into the shaft which drives the eccentric sheave, the later being slotted in its larger half for the reception of the key.

Eccentric Lug. Refers to the projecting portion of an eccentric strap to which the eccentric rod is attached.

Eccentric Rod. Refers to the rod attached to the eccentric strap which transmits the motion of the strap to the valve or pump.

Eccentric Sheave. Refers to an eccentric disc, commonly made in two halves so that it can be placed upon the crankshaft or axle and the portions joined by bolts and cotters, or formed integral with the shaft or directly keyed to the shaft.

Eccentric Strap. Refers to a narrow split bearing, or a metal hoop, which encircles the eccentric sheave or cam and transmits its motion to the valve rods and thus to the valve gearing.

Eccentric Throw-out. Refers to a device by which the back gear shaft of a lathe runs in eccentric-bored bearings to bring the ears in and out of mesh with those on the mendrel.

Eccentricity

    (*a*)   Refers to the half the 'radial run-out' in a gear.

    (*b*)   Refers to the distance from the centre of a figure or revolving body to the axis about which it turns. In an eccentric this distance has been half the throw.

    (*c*)   The deviation from a centre.

    (*d*)   Refers to the perpendicular distance from the centre of application of a load to the centroid of the section of the structural member supporting the load non-axially.

Elastic Constants (moduli of elasticity). These may be expressed in units of stress (N/m^2 – Pascals or ibf/in^2).

Elastic Curve. Refers to the curve or slope of the normally straight flexual axis when a beam gets bent by loads which do not make the stresses in the beam material to exceed the elastic limit.

Elastic Limit. Refers to the limiting value of the force deforming a body beyond which it does not return to its original shape and dimensions after the force gets removed, that is, no permanent deformation. For steels it is usually regarded to concide with the limit of proportionality.

Apparent elastic limit. Refers to the stress level at which the rate of change of strain with respect to stress has been 15% of the value at zero stress in a stress strain plot. It is easier to estimate than the limit of proportionality and is convenient for comparisons of similar materials.

Elastic Ratio. Refers to the ratio of the elastic limit to the ultimate strength of a material.

Elastic Strain
  (a) Refers to a strain in a material which disappears with the removal of the straining force.
  (b) Refers to the amount of such a strain.

Elastic Strength. Refers to the greatest stress which a bar or structure has been capable of sustaining within the elastic limit.

Elasticity. May be defined as the property of a body which returns, or tends to return, the body to its original size or shape after deformation by external forces.

Elasticity, Modulus of. Young's modulus.

Electric Locomotive. A locomotive which is driven solely by electric power.

Electrogyro. A device using the kinetic energy of a large flywheel to power short-haul vehicles, the flywheel being accelerated at charing points along the route

Electrolytic Polishing. Refers to the production of a smooth lustrous metal surface by making it the anode in an electrolytic solution and dissolving the protuberances.

Electronic Control. Refers to the application of electronic techniques to the control of machines and machine tools, power, data and material processing, moving vehicles, aircraft, missiles, sattellites etc.

Electrolytic Machining. Refers to the reverse of electroplating, with the workpiece as the anode and the tool as the cathode.

Elevator
  (a) A lift.

(*b*) A conveyor which is used for raising and lowering material temporarily carried in buckets, etc., and attached to an endless chain.

Elliptic Chuck. A special chuck which finds use in a lathe to enable pieces of material to get turned with elliptical cross- sections.

Elliptic Trammel. An instrument which is used for drawing ellipses, consisting of two straight grooves at right angles to each other and a bar with a pencil at one end and two adjustable studs that are sliding in the grooves.

Elliptical Gears. Toothed wheels elliptical in form, each rotating about a shaft located at a focus of the ellipse. The rotary motion is varying from point to point during the rotation of the wheels. They find use, occasionally for producing a quick return motion on machine tools so as to give a slow cutting and a quick return stroke.

Elongation. Refers to the total extension produced in a test specimen during a tensile test. It is expressed as a percentage of the original length of the specimen.

Emery Surfacer. It is a broad emery wheel, revolving at high speed, under which the work passes on a sliding table.

Emery Wheel. A grinding wheel made of powdered emery cemented by a bonding material.

Emulsified Coolants. Coolants used in metal machining.

End-feed Magazine. A magazine it is used with centreless, grinding machines. It is especially suited to the loading of short cylindrical components.

Encoder. A device which is used for establishing the angular position of a rotating element such as a shaft. The output can be analogue or digital.

End Measuring Instruments. Instruments which are used to measure lengths by contracting both ends of an object, such as micrometers.

End Mill. A milling cutter which having radially disposed cutting teeth on its circular end face for facing operation.

End Quench Test. See Jominy Test.

Endless. A term applicable to a belt, chain or rope to imply that the ends have been joined together so that it can be used for the transmission of power.

Endless Rope Haulage. The haulage of trucks underground, using a long loop or rope guided by pulleys along the roads and actuated by a power-driven winding drum.

Endurance. See limiting range of stress.

Engagement of Screw Thread

(a) *Depth.* Refers to the radial distance by which the thread forms of two mating threads overlap each other.

(b) *Length.* Refers to the axial distance over which two mating threads are designed to make contact.

Engine. A prime mover; a machine in which power has been applied to do work, often in the form of converting heat energy into mechanical work.

Engine (servomotor types). A mechanism which is used for transferring energy from hydraulic fluid to a rotating shaft.

Abutment engine. An engine in which an oscillating piston or a vane (usually double-acting) is rotating a shaft through an angle less than 260°.

Orbital engine. An axial two-stroke engine with curved pistons in a circular cylinder block, the latter is rotating around a fixed shaft.

Swashplate engine. An engine which is constructed similarly to a swashplate sump with a fluid flow input and a rotating shaft output.

Vane engine. An engine which is constructed similarly to a vane pump with fluid under pressure for the input and a rotating shaft output.

Variable-capacity engine. Refers to an engine in which the capacity per revolution can be changed by varying the porting, rotor eccentricity or stroke.

Variable-stroke engine. Refers to a radial engine or swashplate engine having a variable crank throw or swash angle respectively for altering the fluid consumption per revolution.

When the angle gets varied automatically it has been called an 'auto-variable stroke engine'.

Engine Beam. Refers to the beam of a beam engine.

Engine Friction. Refers to the total value for the frictional resistance of all the moving part within an engine.

Engine Indicator. An instrument which is used for recording the pressure in an engine against either a time base or the crank shaft position. Figure given below shows how the pressure, fed to below a piston (shown stippled) in a cylinder, moves the piston upwards against a calibrated spring and is recorded on a chart by a pen to which it is linked. The drum on which the chart has been wrapped oscillates synchronously with the engine piston by a cord linked with the piston motion and kept taut by a spiral spring inside a drum. The pen is making a cyclic diagram on the chart showing cylinder pressure against crankshaft position.

Engine Lathe. See centre lathe.

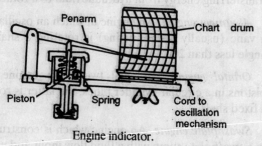

Engine indicator.

Engine Pit (pit)

(*a*)   Refers to a hole in the floor for the inspection of a locomotive or a motor-vehicle.

(*b*)   Refers to the place reserved on a motor-racing track for the inspection, maintenance and refuelling of the racing vehicles.

(*c*)   Refers to the box-like, lower part of the crankcase of an engine; also termed as 'crank-pit'.

(*d*)   A large pit which is used for giving clearance to the flywheel of a large gas-engine or winding-engine.

Engine Power. See brake horse-power; indicated horse-power.

Engine Shaft. See crankshaft.

Engine Speed. May be to defined as the speed in revolutions per minute of the driving shaft of an engine, or of the main rotor assembly in a turbine engine.

Engineers' Blue. A copper-sulphate solution which is prepared by dissolving copper sulphate crystals in water to which a few drops of sulphuric acid are added, or by electrolytic means.

Engrave

(*a*) Cut into metal characters and lines to number of mark parts.

(*b*) Cut into metal lines, shapes, pictures, etc., so that the metal plate may be used to print subsequent copies of the illustration.

Enlarging Drill. Refers to a combination of drill and reamer; a double-fluted twist drill presenting three cutting edges.

Enthalpy. Refers to the heat content per unit mass of a substance.

Entropy. A substance getting a quantity of heat dQ at temperature T is said to have its entropy increased by dQ/T. Entropy is measured in Joules/K or Btu/°F.

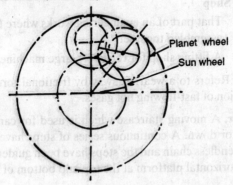

Epicyclic gearing.

Epicyclic Gearing. Refers to a system of gears having one or more wheels travelling round the outside of another wheel whose axis is fixed.

The darker lines in Figure show the relative moment of the sun and planet wheels respectively.

Epicyclic Train. Refers to a system of spicyclic gears in which one or more wheel axes revolve about other fixed axes.

Epitrochoidal Engine. Refers to a rotary type of internal combustion engine having a single central rotating member replacing conventional pistons in cylinders as in the case of a Wankel engine.

Equilibrium. The production of balance as when providing balance weights for a lift.

Equilibrium Ring (valve ring). A ring which is present between the back of the slide valve and the cover in the steam chest of large steam-engines to lessen the amount of friction of the valve by connecting the enclosed space to the exhaust.

Equilibrium Slide Valve. A large slide valve which is balanced by the use of an equilibrium ring.

Equilibrium Valve. A valve in which the pressure on the two sides are made as nearly equal as possible, such as in an equilibrium slide valve and a double-beat valve.

Equipoise. Static balance.

Erecting Shop
- (*a*) That part of an engineering works where finished parts are assembled together.
- (*b*) A fitting shop for relatively large machines.

Erosion. Refers to a wearing away by frictional forces including the action of fast-flowing hot gases.

Esculator. A moving staircase which is used for carrying passengers up or down. A continuous series of steps have been carried on an endless chain and the steps have been guided to flatten out to a horizontal platform at the top and bottom of their run.

Escape Motion. Box of tricks.

Escape Pinion. Refers to the pinion on the arbor of the escape wheel.

Escape Valve. Safety-valve or relief valve.

Evacuator. A vacuum pump.

Even Pitch. Refers to the pitch of a screw thread cut in a lathe which has been equal or a multiple of the pitch of the lathe-head screw (or lead screw).

Excavator. A power-driven machine which is used for digging out and removing earth.

Exhaust

(a) Refers to the working fluid which gets discharged form an engine cylinder after expansion or from the exhaust pipe of a jet-engine or from a steam-engine to the condenser.

(b) Refers to that period of an engine cycle occupied by the discharge of the used fluid.

Exhaust Cone. Refers to an assembly which leads the exhaust gas from the annular turbine discharge to the jet pipe in a turbojet engine.

Exhaust Fan

(a) A fan which is usually placed at the entrance to, or in, a pipe for creating an artificial draught out of a building.

(b) A fan in the smoke uptake of a boiler to exhaust the flue gases.

Exhaust Gas. Refers to the gaseous products of a piston-engine, rocket engine, turbojet engine or other form of gas-driven turbine.

Exhaust Injector. An injector which is operated by the exhaust steam of a steam-engine.

Exhaust Lap (inside lap). Refers to the distance moved by a side valve from the mid-position on the port face before uncovering the steam port to exhaust. It promotes cushioning by closing the exhaust early.

Exhaust Line. Refers to the lower line of the area which is drawn on an indicator diagram. It indicates the back pressure on the piston during the exhaust stroke of a steam or other engine.

Exhaust Pipes. Refers to the pipes through which the exhaust products of an engine get discharged.

Branch pipes. Refers to a short pipe which is able to convey exhaust gases from a piston-engine cylinder to an exhaust manifold.

Ejector pipe. Refers to a exhaust pipe from a piston-engine which gives appreciable forward thrust, as with some aeroplanes.

Stub pipe. Refers to a short pipe which conveys exhaust gases direct from a piston-engine cylinder to the atmosphere.

Tail pipe. Refers to a pipe which leads exhaust gases away from an exhaust manifold.

Exhaust Port. It is the port in acylinder through which a valve allows the escape of exhaust steam or gases.

Exhaust Relief Valve. Refers to a slide valve which opens to a greater width for exhaust than for steam inlet, being fully opened for the exhaust.

Exhaust Silencer. An expansion chamber which is fitted to the exhaust pipe of a piston-engine to reduce the noise level.

Exhaust Stator Blades. See stator blades (exhaust).

Exhaust Stroke (scavenging stroke). Refers to the piston stroke of a reciprocating engine during which the steam or exhaust gases get ejected from the cylinder.

Exhaust System. Refers to the duct or ducts through which the exhaust gases from a combustion system get discharged.

Exhaust Turbine. A turbine which is driven by the exhaust gases of a piston-engine to provide the power for a supercharger, generator or aircraft propeller.

Exhaust Valve. One or more independent valves fitted to those types of engines which are having separate ports for exhausting and admitting.

Exhaust Vanes. Gyroscopically controlled vanes which are fitted in the nozzle of a rocket, for steering while the rocket engine is functioning.

Expandng Bit. A boring bit which is carrying an adjustable cutter on a radial arm for boring holes of different diameters.

Expanding Clutch. Refers to a friction clutch which gets engaged by forcing shoes radially against the inner rim of a disc, cone or drum by the interposition of a toggle joint.

Expanding Ring Clutch. Refers to a form of clutch or coupling, in which the effective agent has been the friction of a split-ring against a bored hole or recess. A metallic ring, divided transversely, has been forced outwards by a wedge against the sides of a parallel-bored hole in the female portion of the clutch, the wedge being actuated by a lever or a screw.

Expansion

(*a*) Refers to the increase in the volume of the working fluid in an engine cylinder.

(*b*) Refers to the piston stroke during the above increase.

(*c*) Refers to the increase in one or more dimensions of a body which is caused by a rise in temperature, or by a decrease of pressure or by ageing.

Expansion Coupling. Refers to a form of coupling in which two thin steel sheets with circumferential corrugations get interposed between two flanged coupling faces which they connect and thus allow a certain amount of end play.

Expansion Curve (line). The line of an indicator diagram which shows the pressure during the expansion.

Expansion Engine. Refers to an engine which uses the expansion of a working fluid.

Expansion Gear. Refers to that part of the valve gear of a steam-engine through which the degree of expansion can get varied.

Expansion Joint. A joint to allow linear expansion or contraction with changes in temperature.

Expansion Ratio. Refers to the ratio between the pressure in the combustion chamber of a rocket or a jet pipe and that at the outlet of the propelling nozzle.

Expansion Rollers. Rollers which are placed underneath the ends of long steel girders, one end fixed, or under the end of a bridge to give support and at the same time allow linear, expansion or contraction with variation of temperature.

Expansion Valve

(*a*) An auxiliary valve which is working on the back of the main slide valve of some steam-engines, to provide an independent control of the point to cut-off.

(*b*)   A regulation valve in refrigeration machines which is used to control the escape of the refrigerant from a liquid state under pressure to a gaseous state.

Expansive Working. Refers to the use of the expansion of a working fluid in an engine, such as when steam, cut off at a fractional part of the stroke, gets caused to do work by its own expansion.

Extensometer. An instrument attached to gauge-points on a test piece. It is used for measuring accurately small strains under load.

Extension Lathe. It is a lathe in which the headstock can be slid longitudinally to provide an adjustable gap of a length to suit the work.

External Screw Thread (external thread). A screw thread which is cut on the outside of a cylindrical bar. Also termed as a 'male thread'.

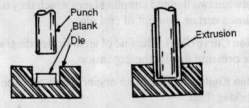

Impact extrusion.

External Screw Tool. A tool which is adapted for cutting external screw threads.

F

F. A symbol for Force.

F. A symbol for signifying stress; also σ.

F. A symbol for Frequency or acceleration.

FHp. Abbreviation for Friction Horse-power.

F/S. Abbreviation of Factor of Safety.

Face Chuck (face plate). A large disc which is used for screwing to the mandrel of a lathe. It is provided with slots and holes for securing work of flat or irregular shape.

Face Cutter. A milling cutter having the teeth radially disposed upon the surface of a disc, either solid with teeth or inserted in grooves.

Face Lathe. A lathe which is used chiefly for surfacing work of large diameter and short length such as large-wheels and discs.

Face-width

    (*a*) Refers to the width over the teeth measured parallel to the axis of a helical, spur or worm gear.

    (*b*) Refers to the width over the teeth measured along the pitch cone generator of a bevel gear.

Face-width, Minimum. Refers to the face-width necessary for ensuring continuity of tooth action in the crossed axis type of helical gear and in worm gears.

    Minimum desirable face-width. The minimum face-width necessary to ensure overlap of tooth action in helical, bevel and hypoid gears.

Facing

    (*a*) Means turning a flat face on a piece of work in a lathe.

    (*b*) Refers to a raised machined surface to which another part has to be attached.

Facing Machine. A centring machine with suitable cutters for facing work.

Factor of Safety. Refers to the ratio of the breaking load on a member, structure or mechanism and the safe permissible load on it. This ratio is permitted when designing the member, bearing in mind the normal conditions of service operation and providing for the possibility of uncertainties of various kinds including variation of strength resulting from deterioration in service.

Proof factor of safety. The factor of safety which is based on the proof load.

Ultimate factor of safety. The factor of safety which is based on the ultimate load.

False Key. A circular key which is driven into a hole parallel with a shaft axis, half drilled in the shaft and half in the hub which gets keyed to it.

Family. A collection of similar parts or mating parts within a sub-assembly or assembly.

Fan

(*a*)   Refers to a rotating paddle wheel or a specially designed propeller for delivering or exhausting large volumes of gas or air with but a low pressure increase.

(*b*)   Refers to a small vane to keep the wheel of a wind pump at right angles to the wind.

(*c*)   Refers to a wheel in a clock mechanism whose velocity is regulated by air resistance.

Fang. Refers to the spike of a tool held in the stock.

Farad (F). The unit of electrical capacitance. It may be defined as the capacitance of a capacitor between the plates of which there has been a potential difference of one volt charged with one coulomb.

Fast Coupling. It is a permanent coupling of two shafts which is consisting of flanges formed integral with the shaft.

Fast Pulley. A pulley which is fixed to a shaft by a key or setscrew.

Fatigue. The process leading to the failure of metals (or other materials) under the repeated action of a cycle of stress. The failure is dependent on the mean stress, the range of stress and the number of cycles. With a decreased amount of stress, a material can be able to withstand a greatly incresed number of repetitions before failure or failure may not occur after millions of stress cycles. With a large amount of stress, failure may take place after a relatively small number of reversals.

Corrosion fatigue. Fatigue which is accelerated by corrosion of the material under stress. For example ferrous metals in a salt water atmosphere.

Fatigue Life. The life of a test-piece, or of a part of a structure or mechanism, expressed as the number of applications of a load before failure.

Safe fatigue life. Refers to the period of time during which the continued applications of a load have been extremely unlikely to result in failure.

Fatigue Limit. The upper limit of the range of stress that a metal is able to withstand indefinitely.

Fatigue Test. Refers to a test on a sample of material or on a complete piece of assembled equipment. The test item has been subject to repetitive loading, usually to verify a safe fatigue life.

Fatigue Testing Machine. A machine which is used for applying rapidly alternating or fluctuating stresses to a test-piece to determinate its fatigue limit.

Faying Face (faying surface). Refers to that portion of a workpiece which is especially prepared to fit a mating part.

Fearnought. Refers to a machine which opens and mixes woollen material, preparatory to carding.

Feather. A parallel key which is partially sunk into a recess in its shaft to allow a wheel to slide axially but not to rotate. It makes machine parts to be thrown into and out of gear.

Feathering (float) Paddles. Refers to paddle wheels which get controlled so that the floats enter and leave the water at right angles to the surface, thereby economizing power.

Feathering Propeller. Refers to a propeller the pitch of which gets altered so that it gives no thrust.

Feed

    (*a*) Refers to the rate at which a cutting tool of a machine gets advanced.

    (*b*) Refers to the advance of material which is operated upon in machine or the provision of materials or requirements necessary for a process or operation.

    Different rates and types of feed have been termed as continuous, fast, intermittent, slow, fine etc.

Feed Box. On preoptive and automatic lathes the feed box gets mounted on the bed below the headstock to provide a range of feeds to both the saddle and slide. The feeds have been independent and reversible.

Feed Finger. A rod which is used with a collect to feed the raw bar forward in a capstan lathe.

Feed Gear. Refers to the mechanism in a machining tool whereby a cutter has been fed to its work or vice versa. A self-acting mechanism.

Feed Mechanism. Refers to the mechanism which is controlling the movements of the carbons of an arc lamp at such a speed as to keep the arc length constant while the carbons burn away.

Feed Pump

    (*a*) A force pump which is used for supplying water for steam boilers.

    (*b*) A force pump which is used for supplying air to gas-engines.

Feed Screw. A screw which is used for controlling the motion of the feed mechanism of a machining tool.

Feedback (back-coupling). Refers to a transfer of energy from an output back into an input. Feedback is considered to be positive when it tends to increase, and negative when it tends to decrease, an amplification.

Feeder. Refers to a machine passes paper, sheet by sheet, into a printing machine.

Felloe (felley). Refers to the circumference or a segment of the circumference of a wheel fitted with spokes.

Female Screw. An internal screw thread.

Ferris Wheel. The giant revolving wheel having a horizontal axis supporting passenger cars which hang freely at its periphery.

Ferrule

(a) A metal band which is forming or strengthening a joint.

(b) A watchmaker's bow drill which is using a small grooved pulley.

Fettle. Means to carve or clear waste material.

Figure-of-eight Callipers. Callipers with jewelled jaws which finds use to hold watch components.

Filament Winding. Refers to the end product or the process of winding fibres on to a mandrel in the formation of cylindrical composite material items.

File Card. A brush having short wire bristles which are used to clean swarf from files.

Filled Rail. A rail which is having one or both sides filled up flush to provide extra strength at points, etc.

Filler Rod (welding rod). Refers to the rod of material which is metal into a joint to complete a connection during welding.

Fillet

(a) A narrow strip of metal which is raised above the general level of a surface.

(b) Refers to the portion of the tooth surface of a gear-wheel which joins the tooth flank and the bottom of the tooth space.

Fillet radius. Refers to the smallest radius of curvature of a fillet.

Film Cooling. Refers to the injection of a coolant directly into a combustion or other chamber, generally through small orifices, so as to provide a cooling film on the inner surface; also called 'transpiration cooling'.

Filter. Refers to a restrictive portion in a fluid flow system which has been designed to remove solid particles. It may be a fine gauze as used in petro and parafin supply lines or papar as for car engine air inlets.

Fin. Refers to one of a number of ribs on the outer surface of a body installed for the dissipation of heat from the body. They have been commonly used on air cooled internal-combustion engines.

Finger

(*a*) A narrow projection or a pin which is used as a guide or guard in a mechanism.

(*b*) A pointer or index.

Finger Guard

(*a*) Refers to any device which serves to guard the fingers of operations from injury by machinery.

(*b*) Refers to protection for a pointer or index of an instrument.

Finger Stop. A sliding stop in a press tool which is used for locating the workpiece under the tool.

Finishing. The term used to describe the final process in the completion of a piece of work such as cutting, rolling, polishing, etc.

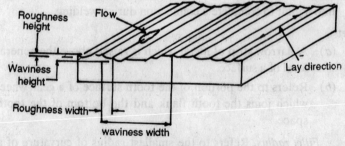

Surface finish.

Finishing Teeth. Teeth which are able to complete the formation of a hole when using a broach.

Finishing Tool. A lathe or planer tool which is used for making a final cut, generally cutting on a wide face.

Fins

(*a*) Refers to thin projecting strips of metal on air-cooled piston engine cylinders to increase the cooling area.

(*b*)   Refers to any thin projecting edges on a metal piece.

Fire Ring. A top piston-ring of a special heat-resisting design which is found in two-stroke compression-ignition engines.

Firing

(*a*)   Refers to the ignition of a charge in a gas- engine.

(*b*)   Refers to the quickening of the source of heat in a steam boiler by adding fuel.

(*c*)   Refers to the overheating of a bearing.

(*d*)   Refers to the ignition of a cartridge starter of a gas turbine engine.

(*e*)   Refers to the operation of a rocket engine.

(*f*)   Refers to the operation of control jets on spacecraft and high-attitude aircraft.

Firing Chamber (lighting chamber). Refers to the small chamber through which a charge of a gas-engine or other engine gets ignited.

Firing Order. Refers to the sequence in which the cylinder of a multi-cylinder piston-engine get fired; for example, 1, 3, 4, 2 for a four-cylinder engine.

Firing Stroke. Refers to the expansion stroke of an internal-combustion engine.

Firing Top-centre. Refers to the top dead-centre before the firing stroke.

Firth Hardometer. A hardness tester which uses a diamond indenter with loads of 10, 30, or 120 kg for hardened steels or a steel ball for soft materials.

Fish Bolt. A bolt which is used for fastening fish-plates and rails together.

Fishplates (splice pieces). Steel cover-plates or cover-straps which are fitted on the side of a fished (butt) joint between successive lengths of rail or beam.

Fit. Refers to the relationship between two mating parts when clearance or interference is present on assembly.

Fitting. Refers to the operations, usually, necessary to complete an assembly other than those done in the foundry and machine shop.

Fittings

    (*a*)   Refers to small auxiliary but essential parts of an engine, machine or mechanism.

    (*b*)   Refers to accessories, especially for boilers, such as values, gauges etc.

Fitting Shop. Refers to the department of an engineering workshop where fitting is done.

Fixed (built-in, clamped, encastre). This describes the way the end of a beam or column or the edge of a plate or shell has been held. It describes the condition where rotation and transverse displacement of the neutral surface has been prevented but longitudinal displacement is allowed along the edge of the plate, for example.

Fixed Cutters. Cutters which are fixed in a machine for the work to be moved over them.

Fixed Eccentric. An eccentric which is permanently keyed to a shaft.

Fixed Expansion. A steam-engine having a constant expansion ratio in which the cut-off cannot be changed.

Fixed Head. Refers to the head of a shaping machine which gets attached to the ram and moves only in the direction of its stroke. Cf. clapper box.

Fixed Headstock (fast head). Refers to the casting bolted to the left-hand end of a lathe bed. It houses the bearings supporting the mandrel and driving pulleys.

Fixed Pulley. A pulley which is keyed to it shaft.

Fixture. It holds the work without controlling the tools.

Flame Cutting. Refers to the cutting of metal with an oxy-acetylene or oxy-hydrogen flame.

Flame Trap. A gauze or wire grid in the intake of a carburetter whose main function is to prevent the emission of the flame from a backfire.

Flame Tube. Refers to the perforated inner tabular 'can' of the combustion chamber of a jet-engine in which the actual burning of the fuel occurs.

Flange

- (*a*) Refers to a projecting rim, as on the rim of a wheel which runs on rails.
- (*b*) Refers to a disc-shaped rim formed on the end of shafts (or pipes) for coupling them together or on to an engine cylinder.
- (*c*) Refers to the top and bottom members of an I- beam.

Flange Coupling. A shaft coupling having two accurately faced flanges which have been keyed to their respective shafts and bolted together.

Flanged. Provided with a projecting flat rim, collar or rib.

Flanged Chuck. Face chuck.

Flanged Nut. A nut having a broad flange integral with its solid face.

Flanged Pipes. Pipes which are provided with flanges for connecting them together by means of bolts.

Flanged Seam. Refers to a joint between two parts around a projecting offset in the workpiece.

Flanged Socket. Refers to a very short pipe with a flange at one end and a socket at the other.

Flanged Spigot. Refers to a very short pipe with a flange at one end and a spigot at the other.

Flanged Wheel. Refers to a wheel having a flange on one or both sides to keep it on the rails.

Flanging Machine (press). A machine (press) which is used for bending over the edges of plates to make a flange.

Flanks

- (*a*) Refers to the curved outlines of teeth in a gear which lie within the pitch circle or below the pitch line.
- (*b*) Refers to the parts of the surface on both sides of a screw thread that intersect an axial plane in straight lines.
- (*c*) Refers to the working faces of cams.

Flanks Clearing Flank. Refers to the flank of a screw thread that does not take the thrust or load in an assembly.

Following flank. The flank of a screw thread which is opposite to the leading flank.

Pressure flank. Refers to the flank that takes the thrust or load in a screw-thread assembly.

Flap. Refers to any hinged or pivoted surface which can get adjusted, usually either automatically or through controls.

Flap-valve. Refers to a non-return valve in the form of a hinged flap or disc which is used for low pressures as found in lifting pumps. It is sometimes faced with leather or rubber.

Forms of flap valves has been ferred 'reflux', 'clapper' and 'swing check' valves.

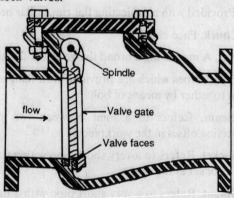

Flap-valve.

Flapping Angle. Refers to the angle between the tip-path plane of the rotor of a rotorcraft and the plane normal to the hub axis.

Flash Point. Refers to the temperature at which a liquid heated in a special apparatus gives off sufficient vapour to flash momentarily on the application of a small flame.

Fleam. Refers to the angle of rake between the cutting edge of a sawtooth and the plane of the blade.

Fletcher's Trolley. A trolley which is designed for the investigation of the motion of a body every part of which has the same velocity at a given moment, and mounted on wheels of small mass.

Flexible Coupling. A shaft coupling which is connecting two shafts, not in rigid alignment, with the drive transmitted through a resilient member like a steel spring, rubber disc bushes or belt and pins.

Flexible Shafting. Refers to a shaft which is consisting of a number of concentric spiral coils of wire, wound alternatley – right and left hand – over each other, thereby giving flexibility when revolving like an ordinary shaft.

Flexural Axis. The line which is joining the flexural centres.

Flexural Rigidity. See flexure.

Flexure. Refers to the bending of a thin bar under forces or moments so that its displacement has been perpendicular to its length. The ratio of the applied force to the displacement has been termed as the 'flexural rigidity'.

Float

(*a*) A small buoyant cylinder which is placed in the float chamber of a carburetter for actuating a valve controlling the petrol supply from a main tank.

(*b*) A buoy which is used to indicate the height of the water in tanks or boilers.

Float Board

(*a*) Refers to the rectangular boards attached to the arms of a paddle wheel.

(*b*) Refers to the boards which receive the impulses of the water in an undershot wheel.

Float Chamber. It is the petrol reservoir in a carburetter from which the jets get supplied for the engine cylinders, and in which the petrol level gets maintained constant by means of a float-controlled valve.

Floating Axle. Refers to an axle on which the shaft gets relieved of all loads or stresses except turning the wheel.

Floating Crane (floating derrick). A large crane which is carried on a pontoon for use in docks.

Floating Mill Wheel. A water wheel having its bearings in a boat moored in a rapidly flowing river. The stream turns the wheel to provide power.

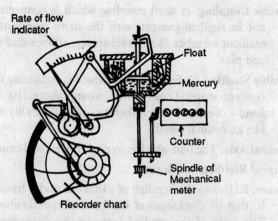

Flow metercentrifugal type registering elements.

Flow-meter, Registering Element. Refers to that part of the meter which makes the rate of flow or total volume past the meter to be recorded. Figure shows a centrifugal type of recording element which provides a visual indication of the flow together with a permanent record. It also stores the integrated value on a counter geared from the meter spindle. With rotation the centrifugal force makes the mercury to flow into the annular chamber; thus the cast-iron float drops, altering the indicator needle and recorder pen positions. A near-linear scaling of the flow-meter has been possible with careful design of the internal bowl shape.

Fluid Drive

(*a*) A constant-torque drive mechanism which is built into the flywheel of an automobile. It consists of two rotors with vanes operating in oil. The amount of slip of the oil between the driving and the driven plate gets varied inversely as the speed, permitting a smooth starting of a car in any gear.

(*b*) A vehicle propulsion system which is used incorporating a fluid flywheel.

Fluid Flywheel. A device which is used for transmitting a rotary drive through the medium of a change in momentum of a fluid (usually oil). The coupling is realised by the action of the fluid on the vanes of the driving and driven tori.

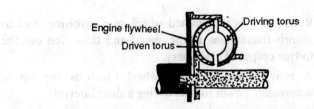

Fluid wheel.

Fluid Lubrication. Lubrication when the bearing surfaces have been completely separated by a viscous oil film, induced and sustained by the relative motion of the surfaces.

Fluid Pressure

(*a*) May be defined as the pressure per unit areas exerted equally in all directions by a fluid. The total force on a horizontal area has been equal to the product of the area, its depth and the density of the fluid.

(*b*) Refers to the pressure transmitted by a fluid.

Fluid Seal. Refers to a seal for a bearing in which a fluid (usually oil) envelops the bearing to prevent the escape of a gas from a high-pressure region to a lower one.

Fluidity. Refers to the inverse of viscosity of loosely, the ability to flow.

Flute. The grooves, straight or spiral, which are running along the length of a twist drill.

Flux. A substance which is added to a solid or applied when the solid gets melted to increase its fusability, by dissolving the oxides which would prevent it adhering to the mating part. It is usual to use a flux when soldering, brazing or galvanizing.

Fly Cutter. Refers to a single-point tool-holder which can be revolved around its axis and by a length adjustment of the horizontal bar in which the tool is mounted a hole can be made of almost any diameter.

Fly Press. A press which is used for punching holes and stamping out thin works in metal.

Fly Shuttle. Refers to the mechanism for propelling a shuttle across a loom.

Fly Wheel

    (*a*) Refers to a heavy-rimmed wheel on a revolving shaft to absorb fluctuations in the speed and thus even out the torque output of machinery.

    (*b*) A heavy-rimmed rotating wheel which is run up to accumulate power for use during a short interval.

Foil. A term which is usually referring to very thin pieces of material, usually metallic and flexible.

Follower

    (*a*) A toothed wheel which is driven by another wheel.

    (*b*) A pinion which is driven by a toothed wheel.

    (*c*) That part of a mechanism, such as a lever arm, which is driven by a cam and usually returned by a spring.

Following Steady. Refers to a steady attached to the back of the side rest of a lathe, which embaces the work behind the tool and follows it along with the rest.

Foot. Refers to a projection on a casting or a forging by which it has been supported or attached to its neighbour.

Foot Brake. A pedal which is operating the brake shoes on wheels of a vehicle, hydraulically or through levers and cables.

Foot Lathe. A light lathe which is driven from a treadle and crank by the foot.

Foot-Iron. Refers to the work done in raising a mass of one ton through a vertical distance of one foot against gravity, equal to 240 foot-pounds.

Foot Valve

    (*a*) Refers to the suction valve of a pump.

    (*b*) A non-return valve at the inlet end of a suction pipe.

Footstep Bearing (footstep). A thrust bearing which is used to support the lower end of a vertical shaft.

Fopple Card. A geometrical device for measuring small amplitude vibrations.

Force. May be defined as the action of one body on another which will cause the second to accelerate unless it is itself acted on by an equal and opposite force. It is vector quantity.

Forced Feed

(a) Refers to the lubrication of an engine by forcing oil to the main bearings and through the hollow crankshaft to the big-end bearing.

(b) The use of a motor-driven pump in a central heating system.

Force Pump (plunger pump)

(a) A pump which consists of a barrel with a solid plunger and valve chest with suction and delivery valve, which delivers liquid at a pressure greater than its suction pressure.

(b) An air pump which is used to clean out gas and other service pipes by blowing air through them.

Forced-circulation Boiler. Steam boilers in which water and steam have been continuously circulated over the heating surface by pumps (as opposed to natural circulation) so as to increase the steaming capacity. Velox and Loffler boilers have been forced-circulation boilers.

Forced Draught. Refers to an air supply to a furnace with the aid of fans or steam jets (as opposed to natual draught created by a chimney) for increasing the rate of combustion.

Forced Lubrication. Means the supply of oil under pressure for the lubrication of engine bearings and of machine tools.

Fore Carriage. Refers to the bogie under the two front wheels of a portable engine.

Forge

(a) A plant where forging is done.

(b) A mill where the rolling of puddling bar is done.

Forge Dics. Shaped metal discs which are used in forging.

Forge Rolls. The train of rolls by which slabs and blooms get converted into puddled bars.

Forge Train. Refers to the series of rolls for rolling out shingled bloom after leaving the steam hammer.

Forge Test. Rough workshop tests, including bending which are made to check the malleability and ductility of iron and steel.

Forging

    (*a*)   A metal article which is formed as a result of a forging process.

    (*b*)   Refers to the operation of shaping metal parts when hot by means of hammers or presses.

Forging Machines. Power hammers and presses/which are used for forging.

Fork

    (*a*)   When beading in a lathe the rolling tool gets held in a fork by the tool slide.

    (*b*)   Refers to the end of the lever which receives the impulse pin in the lever escapement of a watch or clock.

    (*c*)   A double-pronged clip on a tub or wagon for the haulage rope or chain.

Fork-lift Truck. A vehicle having power-operated prongs which can get raised or lowered at will, for stacking and loading, for transporting and unloading, packages of goods; the last are called pallets.

Form Diameter. Refers to the diameter of the circle from which the involute or a gear is designed.

Form Tool (forming cutter). A cutter having a profile similar to, but not necessarily identical with, the shape and contour needed for the workpiece.

Former. A templet which is used for the cutting of gear-teeth, etc., in copying machines.

Forming. When two or more bends are made simultaneously, the bending operation by plastic deformation of a metal sheet is termed as forming.

    Brake forming. Utilizes standard dies in a brake press which is having a long narrow forming bed : standard presses can handle plates up to thirty feet long.

    Electric discharging forming. A high-energy spark gets discharged between two electrodes immersed in water along

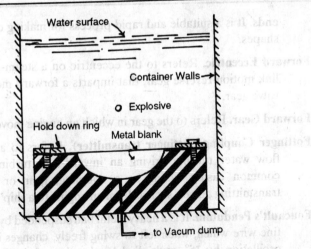

Explosive forming.

with the metal to be formed. The discharge is inducing a shock wave throughout the water which forms the metal instantaneously.

High temperature forming. Refers to the utilization of high energy in the form of heat using either dies integrally heated by cartridge type heaters or hot-fluid forming, in both of which heat and pressure have been supplied by a hot liquid metal alloy.

Rubber pad forming. Forming using a confined rubber pad instead of the conventional mating die, when the resistance offered by the rubber forces the metal to conform to the punch shape.

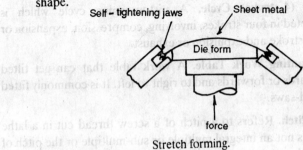

Stretch forming.

Stretch forming. Refers to a process which entails wrapping sheet metal around a male die, pulling and then trimming the ends. It is a suitable and rapid process for making complicated shapes.

Forward Eccentric. Refers to the eccentric on a steam-engine with link motion reverse gear, that imparts a forward motion to the valve gear.

Forward Gear. Refers to the gear in which an engine moves forward.

Fottinger Coupling (fottinger transmitter). Refers to an outward-flow water turbine driving an inward-flow turbine within a common casing, acting as a coupling, gear or clutch for transmitting power, such as from an engine to a ship's propeller.

Foucault's Pendulum. It is a heavy metal ball suspended by a very long fine wire which, when left to swing freely, changes its plane of oscillation by 15°, multiplied by the sine of the latitude, per sidereal hour, thus demonstrating rotation of the earth.

Four-bar Chain. Four links which are connected at their ends by pins to form a trapezium of changeable shape in the plane of the links.

Four-cutter Machine. A machine which is used for cutting four faces of a piece of wood at the same time.

Four-cylinder Engines. Compound steam-engines having two high-pressure and two low-pressure cylinders.

Four-high Mill. A rolling mill which is composed of two small working rolls supported and driven by the larger rolls.

Four-stroke (4-stroke) Cycle. A piston-engine cycle which is completed in four strokes, involving, compression, expansion or power stroke and expulsion or exhaust.

Four-way Canting Work Table. A work table that can get tilted backwards or forwards and to right to left. It is commonly fitted to band-saws.

Fractional Pitch. Refers to a pitch of a screw thread cut in a lathe which is not an integral multiple or sub-multiple or the pitch of the lead screw.

Fracture. Refers to the propagation of a crack through a material, usually due to repeated application of a load.

Fracture Mechanics. Deals with the study of the mechanism of fracture in materials of all kinds.

Frame

    (*a*)   Refers to the structure of the chassis of motor- vehicles.

    (*b*)   Any structure which is built up of compression and tension members.

Frame of Reference. A frame of reference is a specified set of geometric conditions to which other locations, motion or time can be referred. Usually, rectilinear, cylindrical or polar systems or axes have been chosen for the frame of reference but other types can be defined.

Framing

    (*a*)   Refers to the skeleton structure of a locomotive.

    (*b*)   Refers to the vertical adjustment of the picture gate in a cine projector or the adjustment of the picture-repetition frequency in a television receiver to keep the picture stationary on the screen.

Francis Water Turbine. Refers to a reaction turbine type with water flowing radially inwards into guide vanes and on into a runner from which it emerges axially. It is used for heads from about 70 to 500 m.

Frazing Machine. A machine which is used for removing the fin from forged nuts and bolts.

Free-piston Engine or Gas Generator. It is an engine in which the reciprocating portion serves as a gas generator yielding no net mechanical power but supply hot compressed gas at about 770 K (900° F) and 4 bar to the turbine.

Free Turbine. Refers to a separate turbine which is not mechanically connected to the remainder of the engine, as for example the free turbine in some turboprop engines which drives the propeller.

Freewheel. A mechanical one-way clutch, as in an automobile, depending on the wedging action of rollers in the transmission line to transmit torque only when the engine is driving.

Fremont Test. It is a notched bar impact test in which a beam specimen notched with a rectangular groove gets broken by a falling weight.

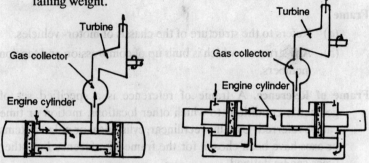

Figure (left). Free-piston gas generator—inward compressor type.
Figure (right). Free-piston gas generator—outward compressor type.

Frequency (periodicity). Refers to the number of vibrations, cycles or waves of a periodic phenomenon per second.

Natural frequency. Refers to the frequency of a free vibration.

Fundamental natural frequency. Refers to the lowest natural frequency.

Undamped frequency. Refers to the frequency of free vibration of a system when undamped.

Angular frequency. Refers to the frequency of a sinusoidal quantity multiplied by 2π.

Anti-resonance frequency. Refers to the frequency at which anti-resonance occurs.

Resonance frequency. Refers to a frequency at which resonance occurs.

Dominant frequencies. Refers to the frequencies of significant maxima in the amplitude spectrum.

Fundamental frequency. Refers to the reciprocal of the period.

Order of frequency. Refers to the number of cycles of an engine vibration occuring during one revolution of the engine shaft, such as, half-order, first-order etc.

Sub-harmonic frequency. Refers to a frequency of which the fundamental frequency has been an integral multiple, such as half-order.

Fretsaw. A very shallow and narrow saw with small teeth held tension in a frame for cutting ornamental patterns and small wooden parts.

Fretting. Wearing away slowly by friction between two surfaces similar to sharpening a cutting tool on an oil stone. Fretting has been an undesirable phenomenon.

Fretting Fatigue (chafing fatigue). Fatigue which is accelerated by fretting.

Friction. Refers to the sliding resistance to the relative motion of two bodies in contact with each other.

    Kinetic friction. Refers to the value of the limiting friction after slipping has occurred, being slightly less than the static friction.

    Limiting friction. Refers to the frictional force, which when increased slightly, will cause slipping.

    Rolling friction. Refers to the frictional force during rolling as distinct from sliding.

    Static friction. Refers to the value of the limiting friction just before slipping occurs.

    Coefficient of friction. May be defined as the ratio of the limiting friction to the normal reaction between the sliding surfaces; the ratio is constant for a given pair of surfaces under normal conditions.

Friction Angle. Refers to the angle that the resultant force makes to the normal to the surface over which a body is sliding when friction is present.

Friction Back Gear. Refers to a lathe device which puts the back gear in and out of engagement by means of a friction clutch, within or next to the cone pulleys, while the lathe is running.

Friction Coupling. A friction clutch.

Friction Disc. A pair of revolving disc having a limited axial movement enclosing a smoothed turned wheel with its axis at right angles.

When contact gets made between the smooth wheel and one or other of the disc wheels, the foremost rotates on its spindle, and by moving one or other of the disc wheels, which rotate in the same direction, the driven wheel and spindle will rotate in opposite directions. This device has been used for reversing the motion of traversing cranes.

Friction Drive. Refers to a drive in which one wheel, pressed into contact with a second wheel, makes the second wheel to rotate by the agency of the friction force between them.

Friction Gear. A gear which is transmitting power from one shaft to another through the tangential friction between a pair of wheels pressed into rolling contact. On contacting surface is generally faced with fabric and the gear is only suitable for small powers. Grooves on the circumferences, which are counterparts of one another, are sometimes provided. The gear uses the same principle as a disc clutch.

Friction Hoist. A light hoist which is driven by the gency of the force between the smooth surfaces of pulleys in contact with each other.

Friction Horse-power (FHp). Refers to that part of the indicated horse-power developed in an engine cylinder which is absorbed in frictional losses, being the difference between the indicated and brake horse-power.

Friction Materials. Cork has been a suitable material for light duty on brakes and clutches, asbestos in a resin binder or rubber compound for moderate pressures and powdered metal on a steel backing for high pressures.

Friction Pulley. A friction wheel having the rim in contact with the rim surface of another pulley.

Friction Ring. A loose metallic ring cut through at one point which gets pressed outwards against a female portion by means of a lever. The device is used in some forms of friction clutches.

Friction Welding (solid phase welding). Refers to the welding of similar or dissimilar metals, the necessary temperature being generated by friction between the two parts being welded. The parts are rotating against one another under an applied load for a given time to form a plastic layer at the interface; then the rotation is stopped. Next, the parts are forged together, without rotation of the materials becoming molten, forming bonds having the strength of the parent metal.

Friction Wheel. Refers to any wheel which drives or is driven by friction, the contact being between smooth or grooved surfaces.

Frog

(*a*) Refers to the point of intersection of the inner rails, in the form of vee, where a train crosses from one set of rails to another.

(*b*) Refers to a metal stop on a power loom to stop its machine whenever, the shuttle is trapped in the warp.

Froude Brake. A hydraulic dynamometer.

Frozen Stress Technique. It is a technique in photo-elasticity where the stresses, which occur under load at room temperature, get retained in the no-load condition by heating the photo-elastic material to about 390 K (243 °F) while under load and cooling down over 10.12 hrs. The material can then be examined and sliced up under no-load conditions at room temperature.

Fuel Tunnel. A device which is used for resetting in flight the automatic fuel regulation of a gas-turbine by barostat to meet changes in ambient pressures.

Fulcrum. Refers to the point of line about which an object balances.

Fulcrum Plate. Refers to the metal plate in an ordinary lift pump which receives the stud about which the handle turns.

Full Gate. Refers to the working of a water turbine when the regulator gets opened so that the whole width of the vanes receives the impact of the water.

Full Gear. When the valves of a steam-engine have been set at the position of maximum travel and cut-off for full power.

Full Thread. A parallel screw thread which is cut to the correct depth for size and pitch.

Fullering

(*a*) Producing circumferential grooves on circular forged work using a 'fullering tool', a tool which consists of a split block internally grooved that is placed round the work and hammered.

(b) Caulking a riveted joint to make it pressure tight.

Full-force Feed. Refers to forced lubrication of an engine, in which the oil goes first to the main and big end bearings and thence by drilled holes or attached pipes to the gudgeon pins and cylinder walls.

Full-load. Refers to the normal maximum load under which an engine or machine is designed to operated continuously.

Fundamental Triangle. It is a triangle with sides representing the form of a screw thread with sharp crest and roots, and pitch and flank angles the same as the basic form; and with the base parallel to a generator of the cylinder, or cone, on which the thread has been formed.

Height (or depth) of the fundamental triangle. Refers to the length of the perpendicular from the apex to the base.

Furrowing. Grooving.

Fusee Chain. A fine-linked chain which connects the fusee to the main-spring barrel.

Fusee Engine. A special lathe which is used for the cutting of fusees.

Fusee Poke. A snail-shaped piece on the smaller end of the fusee.

G

G. Refers to the symbol for Gravitational acceleration.

Gap Lever. The lever which is connecting the slide-valve spindle and the eccentric rod in some marine-engine valves.

Galvanizing. Refers to the coating of steel or iron with zinc by immersion in a path of molten zinc covered with a flux.

Gang

    (a) Refers to a train of mining tubs or trucks.

    (b) Refers to the joining together of two prime movers or locomotives.

    (c) A series of gears or machine tools which are connected together.

Gang Die. A die which is used for a multiple punching machine.

Gang Saws. A number of parallel saws in one frame which are used to cut simultaneously a log into many strips.

Gang Tool. A tool-holder having a number of cutters.

Gantry. The trussed girders in an overhead travelling crane.

Gap Bed. A lathe bed having a gap near the headstock to allow the turning of large flat work of greater radius than the centre height.

Gap Gauge. A pair of anvils which are held in a rigid frame, often of C-shape, to check the dimensions of shafts, external threads, etc. It may be solid for a go or no-go gauge, or adjustable.

Gap Lathe. A lathe which is provided with a gap bed.

Gap Press. Refers to an inclinable 'C'-frame power press which is suitable for work on long or wide components.

Gap Bearing. Refers to a bearing in which air or nitrogen is used as a lubricant; it is introduced through holes in a sleeve surrounding a shaft of either a journal or thrust bearing. Gas bearings find use in low-friction, high-speed applications.

Gas-engine. It is an internal-combustion engine working on the Otto cycle in which gaseous fuel is allowed to mix with air to provide the combustible mixture in the cylinder, the mixture being fired by spark ignition. In the two-port two-cycle engine admission and exhaust take place at the same time; in the three-port engine, a carburetter can be used without a check valve, as the port is opened and closed by the piston.

Gas-engine Starter

(a) A small engine which is used for pumping a gas/air mixture into the cylinder of a large gas-engine.

(b) Refers to a compressed air-supply for starting a large gas-engine.

Gas Exhaust. A large, low-pressure rotary vane pump or a centrifugal blower which is used to exhaust gas from gas work's retorts.

Gas Generator. A gas-producing unit forming a source of power. It is used as a starter for a turbojet engine or to drive an auxiliary power unit in an aircraft. A common type of gas generator has been a compact compressor/annular combustion/single turbine assembly.

Gas Meter. A mechanical device which is used for measuring the amount of gas flowing through a pipe. Figure shows a mechanical gas meter of the semi-positive type which works on the principle of the Roots blower. The two horizontally mounted impeller shafts get geared together at each end. The impeller shape ensures that the two have been always in contact with each other and with the walls via a scraper tip at each end. Gas enters from the top, the pressure differential then makes the impellers to rotate thus allowing the gas to pass down by the outside case. The pressure drop in the meter is proportional to the gas velocity. The flow range has been from approximately 10% of the maximum up to 15% at overload. The accuracy guarantee has been about + 1%.

Gas Meter Wet Type. The gas meter is having a drum which is revolving in a cylinder which is more than half filled with water and divided into compartments by partitions. (See Figure). One

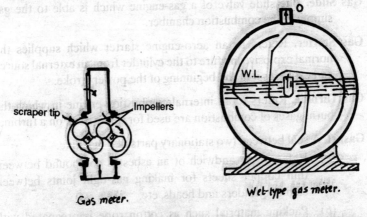

scraper tip — Impellers

Gas meter.　*Wet-type gas meter.*

end of each partition should always be below the water surface. Gas enters by the inlet l and escapes by the outler O, the water level WL being as shown and filling two of the compartments. If the gas gets drawn from one compartment, the pressure of incoming gas rotates the drum and brings another compartment into communication with the outlet. As each compartment fills and empties a revolution indicator attached to the drum would read the delivered volume of the gas.

Gas Ports

(*a*)　Refers to the inlet passages to a gas-engine cylinder.

(*b*)　Refers to the inlet and outlet passages to the cylinders of internal-combustion engines.

(*c*)　A general term which is used for the tubes or pipes leading into a larger volume and usually closed by an inlet valve.

Gas Pump. A small pump which is used for forcing gas into the combustion chamber of some gas-engines.

Gas Regulator

(*a*)　An automatic valve which is used for maintaining a steady gas pressure in gas supply mains.

(*b*)　Refers to the throttle valve of a gas-engine.

(*c*)　Refers to the manually set thermostat in domestic gas-fired equipment.

Gas Ring. A spring ring which is used for maintaining a gas-tight seal between the piston and the cylinder wall.

Gas Slide. The slide valve of a gas-engine which is able to the gas supply to the combustion chamber.

Gas Starter. Refers to an aero-engine starter which supplies the normal explosive mixture to the cylinder from an external source and explodes it at the beginning of the power stroke.

Gas Turbine. Refers to an internal-combustion engine in which the burnt gases of combustion are used for doing work on a turbine.

Gasket. A seal between two stationary parts of a machine.

- (*a*) Refers to a sandwich of an asbestos compound between thin copper sheets for making gas-tight joints between engine cylinders and heads, etc.
- (*b*) Packing material such as cotton rope impregnated with graphite grease for packing stuffing-boxes on pumps, etc.
- (*c*) Refers to any ring or washer of packing material.
- (*d*) A shoft thin metal sheet having ridges which partially flatten on assembly.

Gate

- (*a*) A valve which controls the supply of water in a conduit.
- (*b*) Refers to the annular opening through which the water passes into the vanes of a turbine.
- (*c*) Also refers to a movable barrier to the passage of a fluid or gas, especially sliding and tilting gates.

Gate Valve. The term used for a valve which provides a straight-through passage for the flow of a fluid. The gate gets moved between the body seats by a stem whose axis has been at right angles so that of the body ends which are themselves in line. The actuating thread of the stem is either contained inside the valve or is exterior to the bonnet.

The following have been gate valves: wedge gate valve; sluice valve; double disc gate valve; parallel slide valve.

Gate Wheel. A toothed wheel which is able to control the various gates by which the opening and closing of the ports of an inward flow hydraulic turbine get effected. A key or crank turns a small pinion gearing into the gate wheel.

Gauge

(a) An instrument which is used to determined imensions, capacity, etc.

(b) An accurately dimensioned piece of metal which is used for checking dimensions, such as master gauge and workshop gauge.

(c) Refers to a measuring tool, such as a micrometer gauge.

(d) Refers to the distane between the inner edges or rails or tramways.

(e) Refers to the diameter of wires and rods on some specified schedule.

Gauge Clocks. Small test clocks which are fitted to the outside of a vessel to indicate the liquid level within.

Gauge Glass. A glass tube which is able to provide a visual indication of water level in a tank or boiler.

Gauge Length. Refers to the distance on an external taper, screw, at the pipe end, from the gauge plane to the small end of the screw, measured parallel to the axis.

Gauge Plane. In a taper screw thread, this refers to the plane perpendicular to the axis at which the major cone has the gauge diameter.

Gauges Commonly Used. (dimensions in inches) :

Name	Gauge no			
	0	1	20	36
SWG	0.324	0.300	0.036	0.0076
BWG	0.340	0.300	0.035	0.004
AWG	0.325	0.289	0.032	0.0050
Steel Wire	0.307	0.283	0.035	0.0090
BG (plate)	0.395	0.353	0.039	0.0061

Gear

(a) Any mechanical system which is used for transmitting motion.

(b) Refers to the transmission of rotation by gearwheels.

(c)   Refers to a gear ratio as in transport vehicles, such as first gear, etc.

(d)   The positions of the valve mechanism in a steam-engine such as astern gear, etc.

(e)   Refers to a set of tools foe performing a particular task.

Anti-backlash gear. A double gear which is comprising two gearwheele next to each other as shown by A and B in Figure. One gets fixed rigidly to shaft E while the other gets spring-loaded against the first through spring C. The combination eliminates backlash between D and E. Cf.

Gearbox

(a)   The term used for the complete system of gearwheels for changing the speed from that of an input shaft to that of an input shaft or changing the direction of rotation, or changing the actual direction of a shaft with or without a speed change.

(b)   The box having the system of gearwheels.

Gear Cluster. Refers to a set of gearwheels integral with, or permanently attached to, a shaft such as the lay shaft in the gearbox of an automobile.

Gear Coupling. Involute gear-teeth cut on flanges keyed to the ends of two coaxially aligned shafts mate having internal teeth at the two ends of a surrounding sleeve. This is packed with grease which allows the transmission of torque between the shafts.

Gear-cutters. Milling cutters, nobs, etc., having the correct tooth form for cutting teeth on gearwheels.

Gear-grinding Machine. A machine which is used for grinding a gear to remove slight distortion after heat-treatment, which uses either a formed wheel which fits the space between two gear-teeth or two flat wheels.

Gear Lever. A lever which is used to move gear wheels into or out of engagement probably via sliding pinions.

Gear Miller. A milling machine which is used for cutting gear- teeth with a milling cutter of the correct shape.

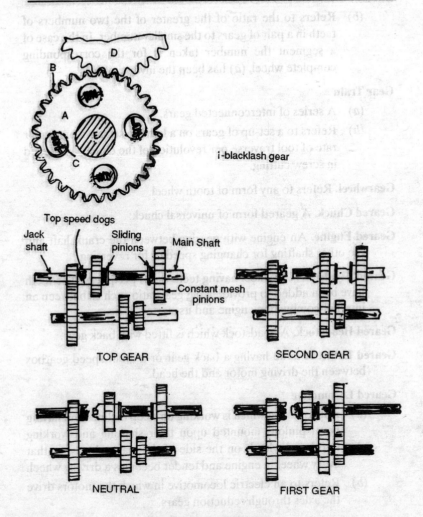

i-blacklash gear

Gearbox operation.

Gear Pump. A small pump which is used for lubricating systems and
the like. It delivers fluid through the tooth spaces of a pair of
gearwheels in mesh and enclosed in a box.

Gear Ratio

(*a*)   Refers to the ratio of the rotational speeds of the input and
output shafts of a gearbox.

(*b*)   Refers to the ratio of the greater of the two numbers of teeth in a pair of gears to the smaller number. In the case of a segment the number taken is for the corresponding complete wheel, (*a*) has been the inverse of (*b*).

Gear Train

(*a*)   A series of interconnected gears.

(*b*)   Refers to a set-up of gears on a lathe to secure a particular rate of tool traverse per revolution of the chuck, as needed in screw cutting.

Gearwheel. Refers to any form of tooth wheel.

Geared Chuck. A geared form of universal chuck.

Geared Engine. An engine with gearing between the crankshaft and the other shafting for changing speed or for reversing.

Geared Flywheel. A flywheel having teeth on its periphery. The teeth have been added to provide a big gear ratio such as between an internal-combustion engine and its starter.

Geared Headstock. A headstock which is fitted with back gear.

Geared Lathe. A lathe having a back gear or a multi-speed gearbox between the driving motor and the head.

Geared Locomotive

(*a*)   A locomotive which is working on steep inclines and having bevel pinions mounted upon their shafting and working into bevel wheels on the side of the main wheels, so that every wheel in engine and tender becomes a driving wheel.

(*b*)   Refers to an electric locomotive in which the motors drive the axles through reduction gears.

Geared Pump. A power pump or series of pumps which is driven by a source of power through spur gearing.

Gearing. A system of gearwheels which is transmitting power.

Gearing-down. Refers to a speed reduction from the driving wheel (or unit) to the driven wheel (or unit).

Gearing-up. Refers to a speed increase from the driving wheel (or unit to the driven wheel (or unit).

General Gauge. A gauge which is designed to serve, under suitable tolerance limits, either as a workshop gauge or as an inspection gauge.

Generator. An apparatus which is used for producing gases, steam, electricity etc.

Gib

(*a*) It is a metal which transmits the thrust of a wedge or cotter, as in some connecting-rod bearings.

(*b*) A brass bearing surface let into the working face of a steam-engine crosshead.

Gibbet. Refers to the triangular framework of a crane which is consisting of post, jib and strut.

Gib-headed Key. A key having a head formed at right angles to its length for securing a wheel, etc., to a shaft.

Gills. Controllable flaps which are used for varying the outlet area of the airflow through a radiator or from the cowing of an air-cooled engine. Also termed as 'radiator flaps' and 'cowl flaps' respectively.

Gin

(*a*) A small hoist which consists of a chain or rope barrel supported in bearings and turned by a crank.

(*b*) A portable tripod carrying lifting tackle.

Gin Block (monkey wheel, whip gin). Refers to a single sheave pulley of hollow-rim section having its bearings in a skeleton frame suspended from a hook.

Gin Pulley. The pulley of a gin block.

Gland

(*a*) A device which is used to prevent leakage at a point where a shaft emerges from a vessel containing fluid under pressure or from a vacuum.

(*b*) A sleeve or nut of one-piece or two-piece design which retains and forms a means of compressing the packing in a stuffing box.

Screwed gland. A gland which gets adjusted by a special nut, the gland nut, to engage with a stuffing box.

Gland packing. Material which is inserted into a gland to disallow leakage of fluid.

Glazing. Means the filling of the surface of a grinding wheel with minute abraded particles so that it gets smooth and polished and no longer grinds efficiently.

Globe Valve. A screw-down valve having the casing or body of a spherical shape. The axis of the stem has been at right angles to the body ends which have been in line with each other. Cf. oblique valve and angle valve.

Glow Plug

    (*a*)   Refers to an electrical igniting plug for switching on to ensure the automatic re-lighting of a gas turbine if the flame becomes unstable, as under icing conditions.

    (*b*)   A heater plug which is installed in each combustion chamber of some diesel engines. The heaters get switched on prior to cold starting and are switched off after the engine has started.

Go and No-go Gauges. Gauges which are used to measure the maximum and minimum metal limits of the work and usually for each independent dimension on the work. Alternatively, 'go and not-go gauges'.

Going Fusee. A fusee with maintaining power.

Gooseneck (tool)

    (*a*)   A tool which is used for giving a finishing cut. The body of the tool is having a semi-circular portion as part of the stem.

Gooseneck, or return, flanging die.

(b) A special press die which is used for forming flanges in sheet metal.

Gorge. Refers to the groove of a sheave pulley in which the rope or chain runs.

Governor. Refers to a mechanism for governing speed by centrifugal force or by pressure, whereby, commonly, heavy balls rotating by the motion of an engine move outwards at higher speeds, and inwards at lower speeds, to close or open a valve controlling the fuel, steam, water or hydraulic supply.

Governor Spindle. Refers to the vertical spindle which carries the governor sleeve and around which the sleeve revolves.

Governor Valve Gear. Refers to the governor-controlled valve gear which regulates the opening and the closing of the induction valve in an automatic expansion.

Grab (grab bucket). A steel bucket which is made of two halves hinged together so that they dig out and enclose part of the material on which they rest. They find use in mechanical excavators and dredgers.

Grab Dredger. A grab suspended from the head of a crane's jib, with the crane raising and lowering the grab. Also known as 'grapple dredger'.

Grabbing Crane. An excavator which consists of a crane carrying a large grab or bucket in the form of a pair of half-scoops, so hinged as to scoop or dig into the earth as they are lifted.

Grain Rolls. Rolls which are made of a tough quality of cast-iron, not chilled.

Grating. The perforated plate in a foot-valve or air-pump which is used for filtering out solid matters.

Gravity Conveyor. Refers to a conveyor in which the weight of the articles has been sufficient to ensure their transport from a higher to a lower level, such as sliding down an inclined runaway.

Gravity Plane. Refers to an inclined plane on which descending full trucks pull up the ascending empty ones.

Gravity Wheel. It is a water wheel in which the weight of the water alone gets utilized. Overshot sheels are gravity wheels.

Grease Box. It is two the upper portion of an axle-box having the grease used for lubrication.

Grease Cock. A cup having pipe and stop-cock screwed into an engine cylinder or bearing housing which is used to receive and regulate the supply of grease for lubricating the moving part.

Grease Cup. A cylindrical cup which is threaded internally an filled with grease with a r'pple screwed into a bearer housing, thereby feeding grease into the bearing when the cup gets screwed down hard against the grease in the cylindrical reservoir.

Grease Gun. A cylinder, filled with grease, from which the grease gets delivered by hand pressure on a piston, intensified by a second plunger which is forming the delivery pipe and gets pressed against a nipple screwed into the bearing that needs lubricating.

Gridiron Valve. A slide valve having the ports in the valve and in the cylinder face subdivided transversely by narrow bars. The valves may be double or treble ported.

Grinder

　(*a*)　A machine tool which is used for shaping pieces to exact size by means of rotating discs of emery, diamond or other abrasive material.

　(*b*)　An emery wheel for grinding tools.

　(*c*)　A large thick circular grit-stone which is used in the manufacture of mechanical wood pulp.

Grinding. Refers to the abrasion of metal surfaces by emery wheels, lapping, etc.

Grinding Clamps. A divided and adjustable lap which is used for grinding mandrels and cylindrical holes.

Grinding Machine. It is a machine tool in which flat, cylindrical or other surfaces get finished by the abrasive action of a high-speed grinding wheel.

Grinding Medium. The solids, such as pebbles or steel balls, used in special mills for grinding materials to powder.

Grinding Wheel. Refers to a wheel which is composed of an abrasive powder such as carborundum or emery, cemented with a binding agent, and fused for cutting and finishing metal.

Grip Chuck. A lathe chuck having movable jaws.

Gripper Feed Mechanism. Refers to a mechanism that grip material, feeding it into a machine, and returns with grippers open.

Grit. The grit, or grain, of a grinding wheel refers to the size of the abrasive particles which are having standard numbers defining the number of meshes to the inch through which the grit will pass.

Grommet. A bung with a hole through it. It is usually made of a non-metallic material and inserted into sheet metal to permit a cable to pass through without abrasion against any sharp edges.

Grooving (Furrowing). Refers to the recessing, by revolving cutters or by a small thick circular saw with widely spaced teeth, of the edges of boards to receive the tongues of corresponding boards.

Grooving Saw. A circular saw which is used for cutting grooves.

Ground Resonance. Rapidly increasing oscillations of the rotor of a rotor craft when run up on the ground, due to the reaction between the dynamic frequency of the rotor and the natural frequency of the alighting gear.

Ground Wheels

(*a*)   Refers to the travelling wheels of a portable crane.

(*b*)   Refers to the wheels attached to seaplane hulls for maintenance when on land, instead of using trolleys.

Ground-off Saw. A saw having a very thin edge almost parallel for about 49 mm (approx. 1½ in), after which one side gets ground with a slightly concave taper until it runs out, leaving thick portion of the plate in the centre of the saw.

Grub Screw. In general a screw having no head but with a slot across the top end for the insertion of a screwdriver. Various types are shown in Figure.

Grummel Washer (grommet washer)

(*a*)   A washer which is made of spun yarn or tar twine, etc. It is used to make a watertight joint under the head of a square-shouldered bolt.

(*b*)   Refers to a hollow rubber or platic washer which is fitted in a hole to permit electric cables. etc., to pass through a hole without chafing.

Hexagon socket   Fluted   Square head   Slotted headless

Flat   cone   Full dog   Half dog   Cup   Oval

Group of setscrews.

Guard Plate

(*a*) Refers to a fixed sheet-steel plate which is in front of machinery to protect pesonnel.

(*b*) Refers to a curved plate in a rubber disc valve which is used to limit the movement of the disc.

Gudgeon

(*a*) The term used for a pivot at the end of a beam or axle.

(*b*) A cross-shaft at right angles to a piston rod or pump rod which is connecting the rod and the crosshead.

Gudgeon Pin (gudgeon wrist pin). The pin which is connecting the piston of an internal-combustion engine with the bearing of the little end of the connecting rod. Also termed as 'piston pin'.

Guide. A chisel-shaped attachement to a rolling mill to lift the rail or bar off the rolls on the leaving side and thereby prevent it from turning around the roll.

Guide Bars. Bars with flat or cylindrical faces which are used for guiding the crosshead of a steam-engine and thus avoiding a lateral thrust on the piston rod.

Guide Blades. Refers to the fixed vanes within a turbine or a compressor which are able to direct the fluid at the proper angle on to the rotating vanes.

Guide Pulley. A loose pulley or idler which is used to guide the direction of a driving-belt wire or cable or to disallow the belt from coming into contact with an obstruction.

Guide Screw Stock. Refers to a die-stock in which the dies have been divided into three portions, one being the guide and the other two the actual cutters which are kept in a radial slots.

Guide Vanes. Vanes which are similar in shape to aerofoils. These guide the airflow in a duct or wind tunnel.

Guillotine Shears. A shearing machine having the shears parallel with the plane of the machine framework. It is used for the cutting up of puddled bars and slabs.

Gullet. Refers to the depression or gap cut in the face of a saw in front of each tooth, alternately on one side of the blade and then on the other.

Gullet Saw (brier-tooth saw). A saw having gullets cut in front of each tooth.

Gullet Tooth. A saw tooth having a gullet cut away in front of it.

Gulleting Machine. A machine which is used for grinding gullets in front of the teeth of a saw.

Gun Drill. A trepanning drill. It removes material at the periphery before the centre. It is used for deep holes.

Gusset (gusset plate). A flat plate which is used to stiffen the joints of a framework; frequently used in conjunction with angle-iron structures.

Guy Derrick. A crane operating from a mast which gets held in an upright position by guy-ropes.

Gymbals

(*a*) A mechanical frame having two mutually perpendicular axes of rotation.

(*b*) Self-aligning bearings for supporting, and keeping level, a chrono-meter in its box.

Gyro (gyro compass). Refers to a type of compass which relies on the principles of rigidity of a gyroscope for its direction- finding ability. It consists of a spinning rotor on flywheel, driven at very high speed by an electric motor or by an air blast, with the axis of spin usually horizontal and mounted in almost frictionless bearings in a double casing. The casing permits freedom of movement about the spinning, horizontal and vertical axes so that the rotor acts as a completely free gyroscope and thereby tends to maintain its direction in shape.

Gyro integrator. Refers to a device in which the total angle of procession of a gyroscope has been a measure of the time-integral of the input torque with the gyro processing at a rate proportional to the torque applied to its gymbal.

Rate gyro. A gyro device which is used for mesuring the rate of change of direction of an axis in space and mounted in a single spring-restrained gymbal. A torque gets exerted on the gymbal proportional to the rate of rotation of the unit in space about an axis at right angles to both the spin and gymbal axes.

Gyrodine. A rotorcraft in which the rotors has been power-driven for take-off, climb, hovering or landing but autorotate (like an autogyro) for crasing flight. The aircraft is having usually short span wings.

Gyroplane. A rotorcraft having a freely rotating rotor.

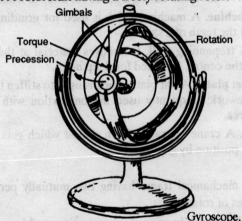

Gyroscope.

Gyroscope (gyro). A small wheel which is rotated at very high speed, generally electrically, in anti-friction bearings. Any alteration in the inclination of the axis of rotation gets resisted by a powerful turning movement (gyrostatic moment) so that the axis remains in a constant direction in space, a property which is used in the gyro compass and the artificial horizon for guiding aircraft, ships, torpedoes and guided missiles.

Gyrostat. Gyroscope.

Half Shroud. A gearwheel shroud which is extending only up to half

H

H-girder. Also known as rolled steel joist or I-beam.

HBN. Brinell Hardness Number.

Hp. Horse-power.

HI. Horizontal Interval.

Hp. High-pressure.

HUCR. Highest Useful Compression Ratio.

Hacksaw

    (a)   A hand-saw which is used for cutting metal. It consists of a
steel frame across which is stretched a narrow saw-blade of
hardened steel.

    (b)   A larger similar reciprocating saw, power-driven through a
crank and connecting rod, usually called a 'hacksawing
machine'.

Hackworth Valvegear. Refers to a radial gear in which an eccentric
opposite the crank operates a link whose end slides along an
inclined guide, the valve rod being pivoted to a point on the link.

Haigh Fatigue-testing Machine. It is a fatigue testing machine in
which the test-piece gets loaded by a powerful electromagnet
excited by an alternating electric current.

Half-centre. Refers to the position of the crank pin of an engine
midway between the two centres.

Half-lap Coupling. Refers to the connection of two co-axial shafts by
a half-lap joint, the two shafts being either riveted together or
enclosed in a keyed-on sleeve.

Half-lap Joint. A joint which is formed by the process of halving.

Half-rip Saw. A hand-saw which is used for cutting timber along the
grain with smaller teeth than a rip-saw.

Half Shroud. A gearwheel shroud which is extending only up to half the height of a tooth.

Half-speed Shaft. Refers to the camshaft of a piston engine which runs at half speed of the crankshaft.

Hammer

(a)   An instrument which is used for beating, etc., or a metal block used for a similar purpose.

(b)   Refers to the weight or weighted mass, which strikes the bells, gongs, tubes or rods in striking and chiming clocks.

(c)   The noise created when a pressure wave travelling along a pipe gets suddenly obstructed. Water hammer in domestic piping has been a common form.

Ball peen hammer. A hammer having a ball-shaped end on the head at the opposite ends of the usual flat striking face.

Cross peen and straight peen hammer. Hammers having blunt chiselshaped ends on the head across on in line with the handle.

Pin hammer. Small light-weight ball peen or cross peen hammers which are used for the lightest work.

Sledge hammer (maul). Refers to a heavy double-faced hammer which gets swung with two hands and may weigh up to 15 kgs (30 lb).

Hammer Blow. Refers to the blow which is caused by the alternating force between the driving wheels of a locomotive and the rails, caused the centrifugal force of the balance weights that balance the reciprocating masses.

Hammer Mill. A mill in which bars, hinged to discs attached to a horizontal rotating shaft, strike the material enclosed in a cage until it becomes fine enough to drop through the bottom openings. It finds use for soft materials like coal and foodstuffs for animals.

Hammer Tongs. In toolmaking, tongs which bend at right angles. These are used to hold to work in making hammers and similar work by hand.

Hand Drill

(a)   It is a tool with a Jacob chuck to take twist drills and powered by hand.

(b) A hand-held electrically powered drilling machine.

Hand Expansion Gear. Variable expansion gear which is usually consisting of right-and left-handed screws for adjusting by hand the positions of a pair of cut-off valves relative to the slide valve.

Hand Feed. Refers to the hand operation of the feed mechanism of a machine tool.

Hand Lift. A lift which consists of a sheave over which an endless rope passes and thus actuates a worm and suitable gearing, worked by hand.

Hand Rest. A support, shaped like the letter 'T'. It is used for resting a hand wood-working or metal spining tool on a lathe.

Handle. A bar shaped to give a good hand grip.

Hanging. The fixing of a pulley etc., upon its appropriate shaft.

Hard Automation. Automation specifically for a particular application.

Hardenability. Refers to the response of a metal to quenching to improve its hardness. The effectiveness has been frequently assesed by a joining test.

Hardness. Resistance to deformation which is usually measured by the resistance to identation by one of various hardness tests.

Hardness Numbers (hardness scale). Refers to any arbitrary scale of numbers which are determined by one of various hardness tests.

Hardness Scale. Hardness numbers.

Hardness Tests. Test which are determined either by (1) the ability of one solid to scratch another (see scratch hardness), or (2) the area of identation formed in a given test (see indentation hardness). Dynamic tests are made to measure rebound hardness by a Herbert pendulum or Shore scleroscope.

Hardness. Refers to a complete set of wiring which are attached at suitable places so that the bundles of wires fit snugly over the finished article and terminate at the appropriate connectors.

Refers to a part of an assembly which surrounds other parts and maintains their relative locations.

Hardness Cord. Varnished linen twine which is connecting the figuring hooks with the mails that lift the warp thread in a jacuard loom.

Hartnell Governor. A spring-loaded governor having vertical arms or bell-crank lever supporting heavy balls and with horizontal arms carrying rollers which push against the central spring-loaded sleeve operating the governing mechanism.

Harvester (combine harvester). A machine which is used for harvesting grain. It cuts the standing cereal crop, separates the grain from straw and chaff depositing the latter in a neat line on the ground behind. The grain gets held in a tank that can be discharged when convenient into an adjacent vehicle even while the harvesting is proceeding.

Harvester-thresher. It is a combine harvester.

Hat-leather Packing. It is a leather packing ring of L-section which is gripped between disc to form a piston or attached to the ram of a hydraulic machine to prevent leakage.

Hawse Pipe (hawser pipe)

(*a*)   A pipe which is able to guide the barrel chain in some types of grabs.

(*b*)   A tubular casting in a ship's bow through which the anchor chain or cable is passing.

Hay Loader. A machine worked by its land wheels picks up hay and conveys it to the towing wagon by means of a trough and reciprocating rake-bars.

Hay Stacker. A machine which is used for catapulting hay up on to a rick.

Head

(*a*)   The height of a liquid column and the pressure resulting from that height.

(*b*)   The top part of a rail.

(*c*)   The driven part of a screw or bolt.

(*d*)   The removable part of an internal-combustion engine above the cylinder or the top of a cylinder.

Head Valve. The delivery valve of a pump.

Header. Refers to a manifold which is supplying fluid to a number of tubes or passages, or connecting them in parallel.

Heading Machine. A power press which is used for producing the heads of bolts, rivets and spikes.

Headstock. A device for supporting the head of a machine like the fixed or poppet head of a lathe, milling machine or grinding machine; the movable head is termed as tailstock, 'movable' or 'loose headstock' and the live head, the 'fixed headstock'.

Headstock Motor. A motor which is used to drive the headstock of a lathe.

Heart Cam

(a) A heart-shaped cam which is used for the conversion of rotary into rectilinear motion.

(b) A heart-shaped cam which is used for deriving an oscillatory motion about a pivot not coincident with the cam axis.

Heat Engine. Any kind of engine that is able to convert heat into mechanical energy.

Heat Exchanger. A device which is used for transferring heat from one medium to another often through metal walls, usually to extract heat from a medium flowing between two surfaces.

Heat Liberation Rate. Refers to a measure of steam-boiler performance given in kJ/1h (= 26.8 Btu/ft^{2}h).

Heat Pump. A machine which is used for transferring heat from a lower grade temperature to a region of higher temperature. It extracts heat from the low temperature body using a refrigerant, which is then compressed to transfer the heat to the higher temperature body, and finally the refrigerant passes through an expander to repeat the whole process. The low and high temperature media have been frequently air or water.

Heating of Bearing. Bearings are said to heat when their temperatures get so hot that the friction gets greatly increased, or the axles or the crankshafts stick fast.

Heddle. A heald shaft.

Heel. Refers to the rear end of the cutting edge of the tooth of a saw of fluted drill.

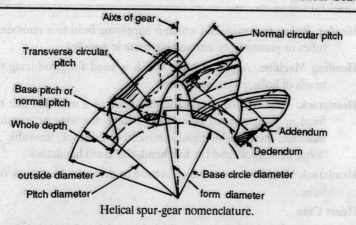

Helical spur-gear nomenclature.

Helical Gear. It is a cylindrical spur gear in which the paths traced by the teeth are helices.

Crossed helical gear. Refers to the helical gears which mesh together on non-parallel axes.

Helical gear-tooth. Refers to that portion of a helical gear which is bounded by the root and tip cylinders and by the two helicoid surfaces.

Helical Spring. A spring which is formed by winding wire into a helix along the surface of a cylinder. It is mainly used to generate tension or compression forces in an axial direction but is sometimes used to produce a torque around the axis when it may be described as a helical torsion spring.

Helicon Gears. Refers to spiroid gear without taper. It can be used at ratios less than 10 to 1. Trade name.

Helicopter. An aircraft having a main lifting rotor or rotor driven by power.

Helve Hammer. A power hammer in which the tup has been secured to a pivoted beam actuated by a crank or cam.

Henry (H). Refers to the unit of electric inductance; it may be defined as the inductance of a closed electrical circuit in which an e.m.f., of one volt is produced when the current varies uniformly at the rate of one ampere per second.

Herbert Pendulum. It is a massive pendulum having a 1.588 mm (1/16 in) diameter steel ball as a pivot which rocks over the surface of a specimen. The period of the pendulum and its rate of damping give a measure of the hardness and ductility of the metal.

Hertz (Hz). The unit of frequency; it is equal to one cycle per second.

High-pressure Cylinder. Refers to the cylinder of a compound steam-engine in which the steam is first expanded.

High-pressure Engine

   (a) Refers to a steam-engine which exhausts directly into the atmosphere.

   (b) A steam-engine which is driven by high-pressure steam.

High-pressure Steam. Steam which is having a pressure considerably higher than atmospheric pressure.

High-speed Steam Engine. Refers to a vertical steam engine, usually compound, with the moving parts of the piston valves totally enclosed and pressure lubricated.

Hitch Feed. An automatic feed mechanism to load strip of sheet material into power presses. It advances the sheet during the forward stroke and holds it there during the return stroke.

Hob. A fluted, straight or helical, rotary cutter which is used to produce spur, helical or worm gears.

Hobbing Machine. A machine using a hob.

Hodograph. A curve which is drawn through the ends of vectors (V_1. V_2.... V_i) radiating from a point (0), the vectors reprecenting the velocities of a body at a successive instants. The velocity (P_1. P_2.... P_i) along the hodograph provides the acceleration of the body along its path.

Hogger Pump. Refers to the upper portion of a deep mine pump.

Hoist. An engine with a drum which is used for winding up a load as in a mine shaft, travelling crane, helicopter etc.

Hole-grinding Machine. It is a machine in which the grinding spindle has been supported plumb with this axis of the work being ground.

Hollander (beating engine). An engine which is used for producing paper pulp. It consists of a trough that contains a beating roll with bars set parallel to the axis of the trough.

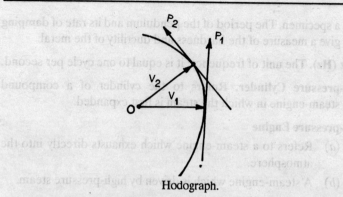

Hodograph.

Hollow Mandrel Lathe. A lathe having a hollow mandrel capable of having bar stock fed through it for repetition work.

Hollo Shaft. A tabular shaft.

Homogenizer. It is a machine that blends or embulsifies a substance by forcing it through fine openings against a hard surface.

Hone (oilstone). A smooth stone which is used either dry or moistened with oil or water, to give a fine keen edge to a cutting tool.

Honey Extractor. Refers to a large cylindrical drum with centre spindle and rotating frame in which honeycombs, after uncapping of the cells, are left for the extraction of the honey by centrifugal force.

Honing. The use of small formed abrasive stone slips mounted in a revolving holder, such as have been used for finishing cylinder bores to a very high degree of accuracy; between 100 and 400 nm (4 and 16μ in) has been the general rule with 25 nm (1μ in) possible. Cf. lapping.

Honing Machine. A machine which is used for finishing cylinder bores, etc., to a very high degree of accuracy.

Hook Tool. A tool which is used for vertical slotting and placed transversely to the axis of the ram of a slotting machine.

Hooke's Joint. Refers to two horseshoe-shaped forks, each pivoted to a single central member carrying two pins at right angles.

Hooke's Law. Strain is proportional to stress in an elastic material below the elastic limit.

Hooke's joint.

Hopper. A container which is used for receiving or feeding supplies of materials from, or to, machines etc.

Hopper Dredger. A dredger having hopper compartments. It is fitted with flap-doors at the bottom, which receive the dredged material and deposit it later elsewhere.

Horizontal Engine. Any engine having a horizontal cylinder axis.

Horizontal Lathe. A boring machine having a vertical axis for boring large engine cylinders and rings.

Horizontal Winch. A steam winch having horizontal cylinders on the side frames.

Horn Balance. Refers to an extension on the outer end of an aeroplane control which gives partial aerodynamic balance and lessens the force needed to move the control.

Hornblock. Refers to a casting which receives the axle box of railway rolling-stock and constrains it to move in a vertical plane.

Horse-power (Hp). It is defined as the engineering unit for power equal to a rate of working of 33000 foot-pounds per minute, 23.56 CHU per minute, 42.42 Btu per minute or 745.700 watts.

Horsepower, Transmitted. Refers to the horse-power from prime movers to mechanisms, via belts, wheels, shafts etc.

Hose. A flexible pipe which has been designed to carry a gas, fluid or slurry. It is used to transmit fluids under high pressure is shown

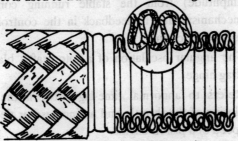

Flexible metal hose (bellows type).

in Figure. The outside has been further protected from damage by external wire braiding.

Hot-air Engine. Refers to an engine in which the working fluid for the heat cycle has been air. There have been many types of hot-air engines, but none have had a high efficiency.

Hot Saw (hot-iron saw). Refers to a metal cutting circular saw which is used for cutting the ends of heated billets, steel forgings etc. The lower portion of the saw runs in cold water.

Hot Well. Refers to the tank or pipes into which the condensate from a steam-engine or turbine condenser gets pumped, and from which it gets returned by the feed pump to the boiler.

Hub. Refers to the central part of a wheel, rotating on or with the axle and from which spokes radiate.

Humidifier

    (*a*) An apparatus which is used for controlling the required humidity conditions in a room for building.

    (*b*) An apparatus which is used for adding moisture to the air in a room, cabin cockpit, space suit etc.

Humpage's Gear. Refers to an epicyclic train of wheels for the speed reduction of a machinery shaft.

Humphrey Gas Pump. Refers to a pump which acts by the periodic explosion of a gas/air mixture above an osciallting column of water in a vessel having an outlet valve. The pump find used in waterworks and gives lifts up to 50 m (150 ft).

Hunting

    (*a*) Refers to an undesired variation (usually of nearly constant amplitude) from the stable running condition of a mechanism due to feedback in the control. Sometimes termed as 'cycling'.

    (*b*) The angular oscillation of a rotorcraft's blade about its drag hinge.

    (*c*) Refers to abnormal time-lag between the opening or closing of the throttle and an increase or decrease in the speed of a piston-engine.

Hunting Tooth. Refers to an extra tooth on a gearwheel so that the number of its teeth shall not be an integral multiple of those in the pinion.

Hydraulic Amplifier. Power amplification which is obtained by the control of the flow of a high pressure liquid by a hydraulically actuated valve mechanism.

Hydraulic Belt. An endless belt of porous material which is driven at high speed with its lower end running under water, which acts like a chain pump.

Hydraulic Brake

(a) Refers to a motor-vehicle brake in which power has been supplied by hydraulic oil pressure via small pistons to expand the brake shoes, the pressure being supplied by a pedal-operated master cylinder and piston.

(b) Refers to a piston in a cylinder filled with some liquid, frequently oil, which absorbs energy through the leakage of the fluid through small holes in the piston of otherwise.

(c) An absorption dynamometer.

Hydraulic Dyanmometer. A dynamometer which is able measure the power being transmitted in a rotating shaft. A rotor attached to the shaft gets mounted inside a casing which has been free to rotate. The space between the two has been filled with water. The change in angular momentum of the water thrown outward by centrifugal force into the stationary pockets in the casing has been measured by the balancing moment of the weight times its distance along the lever arm.

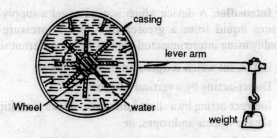

Hydraulic dynamometer.

Hydraulic Efficiency

   (*a*) May be defined as the ratio between the work done by a turbine per pound of water and the available head.

   (*b*) May also be defined as the ratio of the actual lift of a centrifugal pump to the head generated by the pump.

Hydraulic Engine. An engine which is driven by water under pressure from an elevated reservoir or from a loaded accumulator.

Hydraulic jack. A jack with the lifting head carried on a plunger working in a cylinder to which oil (or water) is supplied under pressure from a pump.

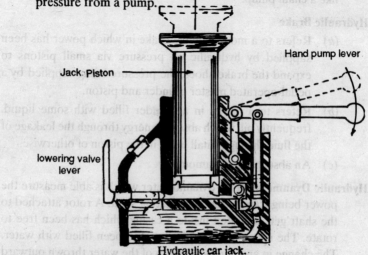

Hydraulic car jack.

Hydraulic Hammer. See hydraulic press.

Hydraulic Hose. A specially manufactured hose for use with high-pressure fluids.

Hydraulic Intensifier. A device which is used to get a supply of high pressure liquid from a greater flow of low pressure liquid, probably using interconnected pistons of two different sizes.

Hydraulic Lift. A lift which is operated by water power.

   (*a*) Direct-acting by a vertical ram, or

   (*b*) Indirect acting by a short ram whose stroke is multiplied by sheave wheels and ropes, or

   (*c*) By adding water to the upper of a pair of lift cars and using a brake on the connecting cable.

Hydraulic Motor. A multi-cylinder reciprocating engine which is driven by water under pressure and usually of radial or swash-plate type.

Hydraulic Piston. A solid piston which is used in force pumps and hydraulic cylinders.

Hydraulic Press. Refers to a ram or piston which is carrying a diehead and working in a cylinder to which high-pressure fluid gets admitted. The work is pressed between this head and a stationary head.

Hydraulic (force) Pump. A force pump which is delivering fluid under high pressure, usually consisting of a number of pistons arranged radially round a crank or operated by a swashplate.

Hydraulic Ram

(*a*) Refers to a mechanism which involves the displacement of a plunger by injecting fluid into, or withdrawing it from, a closed chamber in which the plunger moves through a gland at one end.

(*b*) Refers to a device for using the pressure head of a large moving column of water to deliver some of the water under a greater pressure.

Hydraulic Riveter. A small ram which is operated by hydraulic power to close rivets either directly or through hinged jaws.

Hydraulic Shearing Machine. A shearing machine which is operated by hydraulic power.

Hydraulic Telemotor. Refers to a remote-operating hydraulic mechanism in which fluid displaced by the movement of the input makes a corresponding movement of the output, with systems usually hermetically scaled to yield a positive displacement.

Hydraulic Test. A test for pressure tightness and strength or for fatigue, by pumping water into a vessel up to a prescribed pressure.

Hydraulic Turbine (water turbine). A turbine which is driven by water fluid.

Hydraulically-operated Disc Brake. A disc brake in which the pressure has been applied by hydraulic piston.

Hydrodynamic Governor. It is a small centrifugal pump whose pressure head, which varies with speed, acts on a piston connected to the regulating valve and thus acts as a governor.

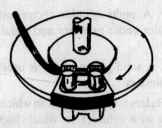

Hydraulically-operated disc brake.

Hydrodynamic Suspension. A suspension system which involves two or more fluid spring elements interconnected so that displacement at one unit is made to affect an adjacent unit. This system is commonly used for reducing the pitching rate of road vehicles.

Hydrostatic Press (bramah's press). A machine which consists of a pair of interconnected cylinders fitted with watertight pistons, the cylinders having different diameters in order to achieve a mechanical advantage.

Hydrostatic Valve. Refers to a valve mechanism that keeps a body, like a torpedo, at a given depth in a fluid.

Hypoid Bevel Gear. It is a bevel gear having the axes of the driving and driven shafts at right angles, but not in the same plane which makes some sliding action between the teeth.

Hypoid Gear. Crossed helical gears which are designed to operate on non-intersecting as well as non-parallel axes, that is, the axis of the pinion is above or below that of the gear and is not parallel to the axis of the wheel.

Hypoid Offset. Refers to the perpendicular distance between the non-intersecting axes of the cones of two hypoid gears.

Hysteresis. Refers to the phenomenon which is exhibited by a system whose state depends on its previous history.

Internal hysteresis. Refers to the property of a material which dissipates internal energy under cyclic deformation.

Hysteresis loop. Refers to the enclosed curve when a cycle of operations get completed, for example, the different stress strain curves when a load gets steadily increased and then decreased, making a loop in the case of some materials.

Stable hysteresis loop. Refers to a hysteresis loop which follows exactly the same path for each and every cycle.

I

I. Refers to the symbol for moment of Inertia.

ICAM. Abbreviation of Integrated Computer Aided Manufacturing.

IcE. Abbreviation of Internal-combustion Engine.

IHp. Abbreviation of Indicated Horse-power.

IMEP. Abbreviation of Indicated Mean Effective Pressure.

I-beam. A structural beam which is having a cross section in the shape of an I. The top and bottom parts take most of the loads applied by bending whereas the centre portion, the web., takes most of the shear load (See Figure). Also termed as H-beam or rolled steel joist.

Ideal Efficiency. Refers to the theoretical maximum efficiency.

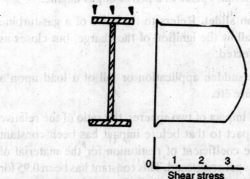

Shear stress distribution in an I-beam section.

Idle Pulley. A pulley which is used in a similar manner to àn idle wheel.

Idle Wheel (carrier wheel, cock wheel)

(*a*) A wheel introduced in a gear train either to reverse rotation or to fill up a gap in the spacing of centres, without affecting the drive ratio. Also known as a 'cock wheel'.

(*b*)   An intermediate wheel.

Idling. Refers to the slow rate of revolution of a piston-engine when the throttle has been in the closed position.

Igniter. Refers to a device by which the charge in a gas-engine or a rocket-engine gets ignited.

Ignition

(*a*)   Refers to the firing of an explosive mixture of gases in an internal-combustion engine by using an electric spark or by a jet of gas in a gas-engine.

(*b*)   Refers to the commencement of combustion in a jet-engine or a rocket-engine.

Ignition Lag. Refers to the time-interval between the passage of the spark and the resulting pressure rise in a cylinder due to combustion.

Ignition Slide. See ignition valve.

Ignition Timing. Refers to the crank angle relative to top dead centre at which the spark takes place in a petrol-or gas-engine.

Ignition Valve (ignition slide). Refers to the valve of a gas-turbine which opens to allow the ignition of the charge, but closes as soon as this is effected.

Impact. Refers to the sudden application or fall of a load upon a specimen, structure etc.

For the direct impact of two spheres, the ratio of the relative velocity after impact to that before impact has been constant and has been the coefficient of restitution for the material of which the spheres get composed. This constant has been 0.95 for glass and 0.2 for lead, the values for other solids lying between these two figures.

Impact Loading, Safe (impact load factor). Refers to the maximum acceleration to which equipment, etc., can get subjected under impact or shock without mechanical damage or operational break-down. The magnitude of the acceleration is given in multiples of g; its duration and its rate of change should be specified.

Impact-testing Machine. A machine which is used for testing the strength of test-specimens under a single blow and for measuring the amount of energy absorbed in a fracture of the specimen. The commonest form of test-piece has been a notched bar.

Impact Wheel. A water wheel which is driven by the impact force of water acting at right-angles to the projecting vanes on the periphery. Turbines have been impact wheels.

Impeller

    (*a*)   Refers to the rotating member of a centrifugal pump compressor, or blower, which imparts kinetic energy to the fluid (water, air, etc.).

    (*b*)   Refers to a rotating member in some meters such as a gas meter.

Impermeator. A form of self-acting lubricator which is used for the cylinders of steam-engines, a two valve contrivance by which steam extracts oil from a small brass cylinder, then condenses to water so that the oil floats to the top and some overflows back to the brass cylinder.

Impulse. It is a vector quantity and is the product of force and time. If the force is not of constant magnitude but is constant in direction, the impulse is $\int_{t1}^{t2} F\, dt$ the time interval $t_2 - t_1$.

Impulse Pin. Refers to the vertical pin in the lever escapement roller which receives the impulses from the pallets, via the notch in the lever, and which effects the unlocking on the reverse vibration.

Impulse Plane. Refers to that part of the pallet upon which a tooth of the escape wheel acts.

Impulse Turbine. It is a steam-turbine in which steam, expanded in nozzles, gets directed to the curved blades carried by rotors, in one or more stages. No change of pressure takes place as the steam passes the blade-ring.

    Combined-impulse turbine. An impulse turbine in which the first stage is consisting of nozzles directing the steam on to two rows of moving blades with a row of fixed blades, the stator in between.

Impulse Wheel. The rotor wheel of an impulse turbine which is similar to the enclosed wheel of a reaction turbine.

In Gear. Mechanisms and engines are said to be 'in gear' when connected ready to be operated, or operating.

Inching. Making adjustments by very small stages.

Inch-pound. Refers to the lifting of a pound weight a distance of one inch. Foot-pound is the more commonly used unit of work.

Inch-ton. Refers to the lifting of a ton weight a height of one inch. Foot-ton is the more commonly used unit of work.

Included Angle. Refers to the angle between the flanks of a screw thread measured in an axial plane.

Increasing Pitch. A screw is regarded to be of 'increasing pitch' when the distance between each successive turn of the helix increases in amount, or the pitch may get increased in the direction of the length of the blade from the centre to the circumferences.

Indentation Hardness. Refers to the estimation of the hardness by the permanent deformation formed in a material by an indenter. The hardness is expressed in terms of the load and the area of the indentation formed. Bell-shaped indenters find use to measure Brinell hardness number and Meyer hardness number. Conical indenters were introduced by Ludwik and the hardness number is equal to the load divided by the surface area of contact between indenter and material. Pyramidal indenters get shaped like a square-based pyramid and are used for determining Vickers hardness numbers, in Rockwell hardness tests, in Knoop hardness tests, and in the Firth hardometer. Dynamic or rebound hardness has been measured by the Shore rebound scleroscope and the Herbert pendulum.

Indenter. An instrument used for making indentations in materials. The depths of identation give a measure of their hardness.

Independent (jaw) Chuck. A chuck for a lathe in which each of the jaws gets moved independently by a key to give very accurate centring for work of irregular shape.

Independent Whip Crane. A plateform crane.

Index Centres. The centres in the headstock and tailstock which are used on milling machines and gear cutters.

Indicated Horse-power (IHp). The power which is shown by an indicator, being that developed by the pressure-volume changes of the working fluid within the cylinder of a reciprocating engine and therefore greater than the brake horse-power by the power lost in friction and pumping.

In a piston engine it may be put as follows : given by ASPNn/ 33000, where A denotes the area of the piston in square inches, S the stroke in feet, N the number of cylinders, n the number of strokes per minute and P is the pressure on the piston in pounds per square inch.

Indicated Mean Effective Pressure (IMP). Refers to the average pressure which is exerted by the working fluid in an engine cylinder throughout the working cycle and equal to the mean height of the indicator diagram in pascals ($1\ Pa = 1\ N/m^2$) or in pounds per square inch.

Indicated Thermal Efficiency. Refers to the ratio of the heat energy equivalent to the indicated horse-power output and the heat energy supplied in the steam or fuel in a reciprocating engine.

Indicator. An instrument which is used for getting the pressure-volume or the pressure-time changes in a steam-engine, or a piston-engine, cylinder during the working cycle.

Indicator Diagram. The diagram drawn by the indicator representing to scale the work done during a cycle of operation of the engine.

Indirect-acting Slide Valve. Refers to a slide valve in a locomotive that gets actuated from an intermediate rocking shaft or double-ended lever, attached at one end to the valve rod and at the other end to the die block of the slot link.

Indirect Action. A motion which is derived through the medium of levers that are not directly related to the motion of the member which is supplying the motive power.

Induced Draught. An artificial draught which gets operated by suction, as with a fan, as opposed to forced draught.

Induction

    (*a*)   Refers to the secondary flow of a gas (or liquid) induced by a primary flow of gas (or liquid).

    (*b*)   Refers to the admission of steam into a cylinder.

(c) Refers to the admission of the explosive mixture into the cylinder or combustion space of an internal-combustion engine.

Induction Port (induction valve, inlet port). Refers to a port or valve through which a charge gets induced into a cylinder during the induction stroke.

Induction Stroke (charging stroke, intake stroke). Refers to the suction stroke during which the working charge, or air, gets induced into the cylinder of an engine.

Inertia. Refers to that property of a body by which it tends to resist a change in its state of rest or of uniform motion in straight line. Inertia may be measured by mass when linear velocities or accelerations are considered and by moment of inertia for rotations about an axis.

Intertia Governor. Refers to a shaft type of centrifugal governor using an eccentrically pivoted weighted arm which responds rapidly to damped speed fluctuations by reason of its inertia.

Ingot. Refers to a metal casting of a suitable shape for subsequent rolling or forging.

Ingot Tilter. Refers to a machine by which ingots are turned between each pass of the rolling mill.

Injection. Refers to the process of injecting fuel into the cylinder of a compression-ignition or petrol engine by using a special pump.

Injection lag. Refers to the time-interval between the beginning of the delivery stroke of the fuel-injection pump of a compression ignition engine and the beginning of the injection of the fuel into the cylinder.

Inlet Valve

(a) The valve in a piston-engine which is used for admitting the steam or fuel-air mixture.

(b) A foot valve.

Inner Dead Centre (top dead centre). Refers to the piston position of a reciprocating engine or pump at the beginning of the out-stroke when the crank pin has been nearest to the cylinder.

Insertion Head. Refers to an automatic feed mechanism which is used for feeding components axially into an assembly, with cutting, clinching and forming tools.

Inside Crank. A crank having two webs which is kept between crankshaft bearings and has the big end of the connecting rod between the two webs.

Inside Cylinders. Locomotive cylinders which have been fixed within the framing and smoke box.

Inside Framing. A form of locomotive framing in which the wheels have been inside the main frames.

Inside Lead (internal lead). It is the degree of opening of the exhaust port of a steam-engine by the slide valve when the piston has been at the bottom dead-centre.

Inspection Gauge. A gauge which is used in the final inspection of a part for testing the accuracy of the finish.

Instantaneous Grip Vice (sudden grip vice). A vice which is operated by levers, a toggle-joint and a rack, instead of a screw.

Instroke. A stroke of a gas-engine piston in a direction towards the ignition chamber.

Intake Stroke. See induction stroke.

Integrator. It is a calculating machine like a computer, or a mechanical machine, such as planimeter, which computes what has been represented mathematically by an integral sign.

Interchangeable Gears. Gears having the teeth of the wheels so designed that other gear wheels of the same diametral pitch, but with any number of teeth, will mesh together correctly.

Internal-combustion Engine. Refers to an engine in which the combustion of a gaseous, liquid or pulverized solid fuel provide heat which gets converted into mechanical work through piston or turbine.

Internal Expanding Brake. See brake shoes.

Internal Feedback. Information from the internal measuring system (resolvers, encoders, tachometers, etc.) which is concerning the motion of a servo-driven robot. It is used for correcting the motion relative to the nominal values defined in a program.

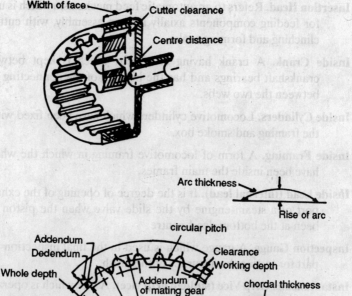

Fig. 2. Internal spur-gear nomenclature.

Internal Spur Gear. A spur wheel having teeth on the inside of the periphery (Figure 1). The nomenclature has been given in Figure 2.

Inverted Cylinder Engine. A vertical engine with the inverted cylinder above the piston rod, connecting-rod and crank.

Inverted Engine. An engine having its cylinders below the crankshaft.

Involute Gear Teeth. Wheel teeth whose flank profile has been the locus of the end of a string uncoiled from a base circle.

Irreversible Transmission. When power can be transmitted, but not reversed.

Iris Diaphgram. A continuously variable hole which forms an adjustable stop for a camera lens and usually integral with the mounting of the lens.

Isentropic. Refers to an isentropic process is one in which there is no change in entropy.

Isochronous Governor. A governor in which the equilibrium speed has been constant for all radii rotation of the balls within the working range.

Isoclinic. Within a stressed body, it refers to the imaginary line along which all points are having corresponding principal stresses with the same orientation.

Isolator. It is a separate mounting, such as a heavy concrete block which is used to isolate a machine or instrument from external vibrations or shock.

Isometric Projection. Refers to an engineering drawing projection are three mutually perpendicular axes which are shown equally inclined to the plane of projection in Figure.

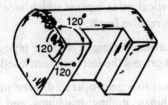

Isometric projection.

Isothermal

(*a*) Taking place at a constant temperature.

(*b*) A line which is joining points at equal temperature.

Isothermal Process. A process such as the compression or expansion of a gas without any change in temperature.

Isotropic Material. A material whose physical properties do not vary with direction.

Izod Test. See notched-bar test.

J

J *(a)* Refers to the symbol for joule, the force of one newton which is acting over the distance of one metre.

(b) Also, refers to the symbol for the polar moment of inertia of a shaft.

Jack

(a) A machine which is used for raising a heavy weight through a short distance by using a screw with gear or hydraulically.

(b) A frame of horizontal bars which is used for supporting fixed vertical wires against which lace bobbins containing yarn can revolve freely.

Jacket. Refers to an outer casing of a boiler, pipe or cylinder having a liquid to heat or cool the internally enclosed object.

Jacob Chuck. Refers to a gear-operated three-jaw chuck for use on drilling machines, milling machines and lathes. It is most frequently seen on portable hand drills. The solid one-piece body gets drilled to fit round the jaws.

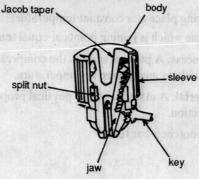

Jacob chuck.

Jacob's Ladder. Refers to a belt conveyor with attached buckets or cups.

Jacot Tool. A watchmaker's tool which is used for polishing and burnishing pivots.

Jacquard Machine. Refers to a weaving machine for operating the shedding and controlling the figuring of a large number of warps in a loom.

Jaw. Refers to that part of a machine which grips hold of the work-piece.

Jerk Pump. Refers to a timed fuel-injection pump with a cam- driven plunger overrunning a spill port to make an abrupt pressure rise which is necessary to initiate injection through the atomizer.

Jet Condenser. A condenser in which steam is condensed by a water spray.

Jet-engine. Refers to a colloquial term for any type of engine which produces thrust by using a jet of hot combustion gases.

Jet Pump. Refers to a pump which delivers large quantities of fluid at a low lift by means of the momentum imparted to the column by the velocity of a jet of steam or compressed air.

Jib. The term used for the inclined or horizontal boom (strut member) of a crane or derrick.

Jib Legs. Refers to legs pivoted to the jib of a breakdown crane and reaching to the ground to provide a firm base for lifting.

Jig. It is an appliance which accuratley guides and locates tools during the operations in a machine shop for producing interchangeable parts.

Jig Borer. Refers to the precision drilling and boring machine which is used in the toolroom for making master jigs and prototype precision work.

Jigger

    (a)   A hydraulic lift which gets operated by a short-stroke hydraulic ram through a system of ropes and pulleys which increase the travel.

    (b)   A potter's wheel.

Jim-crow

    (a)   Refers to a swivelling tool-head, cutting during each stroke of the table of a planning machine.

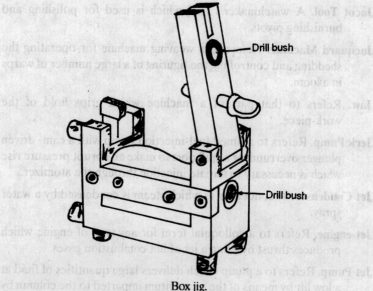

Box jig.

(*b*)   Refers to a rail-bending device.

Jockey Pulley. It is a small pulley wheel which is weighted so that it keeps a drive belt or chain taut as shown in Figure.

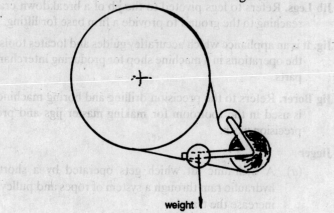

Jockey pully.

Joint Pulley. A pin which connects the two parts of a knuckle joint and continuous surface.

Jominy Test (end-quench test). A test which is used for determining the relative harden ability of steels in which one end of a heated cylindrical specimen gets quenched, the resulting hardness decrease towards the unquenched end given a measure of improved hardness.

Joule. It is a unit of work, energy and heat. It may be defined as the work done when a force of one newton moves through a distance of one metre.

Jounce. Refers to the initial impact of an automobile wheel hitting a raised obstruction on a road surface; the opposite of rebound.

Journal Box. An axle box.

Joy's Valve-gear. A locomotive valve-gear of the radial valve-gear type having no eccentrics, the valve rod being worked directly through a coupling rod or link from the connecting-rod.

Jumper Spring. A spring which is attached to a jumper for holding a star wheel in place.

Jumping-in. Refers to the spring-in of a ring on a piston so that it will enter the cylinder of a piston-engine.

Junk-ring

   (*a*) A metal ring which is attached to a piston of a steam-engine for confining soft packing materials.

   (*b*) Similar to (*a*) for holding a cast-iron piston-ring in position.

   (*c*) A ring which is used for maintaining a gas-tight seal between the cylinder-head and the bore of a sleeve valve.

K

K. Refers to the symbol for the radius of gyration.

Kgf. Refers to a force equal to the weight of one kilogramme.

Kc/s. Abbreviation of kilocycles per second.

Km/ph. Kilometres per hour.

Km/pl. Abbreviation of kilometres per litre.

Kn. Abbreviation of knot.

KWh (kwhr). Abbreviation of kilowatt-hour.

Kaplan Water Turbine. It is a propeller-type water turbine in which the pitch of the blades can be varied in accordance with the load, with a consequent improvement in efficiency.

Karrusel Movement. Refers to a movement in which the revolving carriage, unlike the tourbillon, is carrying no power to the escapement and is carried, not driven, by the third pinion. The fourth pinion would receive power direct from the third wheel and the fourth wheel revolves in the usual way driving the escape pinion.

Kater's Pendulum. A bronze bar having a fixed and a movable knife-edge. The latter gets adjusted until the time of oscillation about either remains the same. The pendulum provides one method of determining the acceleration due to gravity at any place.

Keep Plate. Refers to a plate which is fitted in a fixed shaft to position a rotating part in an axial direction.

Kelvin (K). The kelvin of thermodynamic temperature may be defined as the fraction 1/273.16 of the thermodynanamic temperature of the triple point of water.

Kennedy Water Meter. It is a water meter in which the volume of flowing water continually fills and empties a cylinder of known volume, the discharge being automatically registered. It empties by tipping when the centre of gravity rises beyond a certain point.

Kentledge. Scrap material which is used for counterbalance on a balance crane.

Key

(a) Refers to a piece of iron or steel which is inserted between a shaft and a hub to prevent relative rotation and fitting into a keyway parallel with the shaft axis.

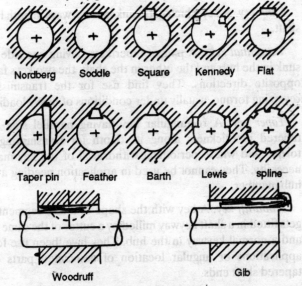

Different types of keys.

(b) A spanner or wrench which is used for tightening the jaws of lathe chucks, etc.

Dovetail key. A parallel key in which the part sunk in the boss has been of dovetail section, the portion on the shaft being of rectangular cross-section.

Gib-headed key. Refers to a key having a head formed at right angles to its length to facilitate withdrawal.

Round (Nordberg) key. Refers to a circular pin or bar which fits into a hole drilled half in the boss and half in the shaft parallel to the shaft axis, usually for light work.

Parallel key. A rectangular key having parallel side which is used in marine tailshafts, etc., when the shafts are greater than 1 in diameter.

Saddle key. A key having a concave face bearing on the surface of the shaft which it grips by friction only, being sunk in a keyway in the boss.

Split key. A key which is split at one end like a split-pin so as to lessen the tendency to work out of its bed.

Sunk key. A key which gets sunk into keyways in both hub and shaft.

Tangential keys. A pair of taper keys having one side largely sunk in the hub and the other in the shaft, the two keys facing in opposite directions. They find use for the transmission of reversing torque, usually under conditions of heavy loading.

Taper key. A rectangular key having parallel side slightly tapered in thickness along its depth for transmitting heavy torque and where periodical withdrawal of the key may be a necessity. They cannot be used in application needing a sliding hub member.

Woodruff key. A key with the shape of a disc segment which gets fitted in a shaft keyway milled by a cutter of the same radius, and a normal keyway in the hub. They have been use for light applications or angular location of associated parts on the tapered shaft ends.

Key Bed. A keyway.

Key Boss. A local thickening of a hub at a point where a keyway has been cut to compensate for loss of strength due to the cut.

Key Chuck. A jaw chuck having jaws adjustable by screws turned by a key spanner.

Key Gauges. Plate gauges which are used for checking the width of keys and key seating.

Keyboard
  (*a*) The banks of keys in an instrument.
  (*b*) A similar arrangement of keys in a typewriter, linotype machine, etc.

Keyway (key-seating). A shallow longitudinal slot which is cut in a shaft or a hub for receiving a key as shown in Fig. 1.

Keyway-seating Machine. A machine tool which is used for milling keyways in shafts, etc., using an end mill with the work supported on a table at right-angles to the tool axis.

Keyway Took (keyway cutter). A slotting machine tool which is used for the vertical cutting of keyways, the tool being equal in width to that of the keyway.

Kibble. A large unguided bucket which is used in shaft-sinking.

Kilogram (kg). It is the unit of mass; it may be defined as equal to the mass of the international prototype of the kilogram.

Kilogram-metre. A unit of work refers to the amount of work done in raising a mass of one kilogram through a vertical distance of one metre against gravity.

Kilowatt-hour (kwh). A unit of energy; may be defined as the energy expanded when a power of 1000 watts is supplied for one hour.

Kinematics. Refers to the study of the velocities and accelerations of the various parts of a mechanism.

Kinematic Chain. A mechanism composed of a closed chain of paired links, the movement of any link being absolutely constrained in relation to all the others in the chain.

Kinematic Pair. Refers to two elements or links which are connected together in such a way that their relative motion is partly or completely constrained.

Kinetic Energy. See energy.

Kinetic Friction. See friction.

King Lever. Refers to a master lever in a signal-lever frame which is able to control the interlocking of the levers to the various railway signals.

Kingpin (swivelpin). A pin joining the stub axle to the axle-beam of an automobile and inclined to the vertical for providing caster action.

Knife-edges. Hardened-steel bearing edges which are working on a horizontal surface on the inner circumference of a ring, allowing fine balance of the adjacent part.

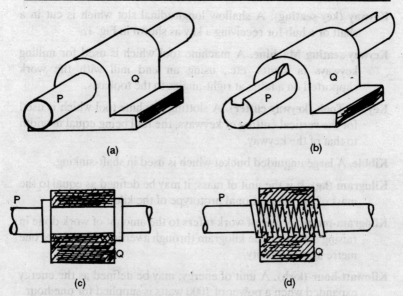

Kinematic pair : (*a*) incomplete restraint, (*b*) (*c*) (*d*) completed restraint,
(*b*) sliding, (*c*) turning, (*d*) screw pair.

Knife Tool. A lathe finishing tool having a straight lateral cutting edge
which is used for turning right up to shoulder or corner.

Knitting Frame. A knitting machine which is using either bearded
needless or latch needles.

Knitting Machine. An apparatus for knitting which is using a single
thread and furnished with a number of hooked and barbed
needles and also loopers for forming the meshes.

Knock Rating. Refers to the percentage of octane in an octane-
heptane mixture of equivalent susceptability to knocking (pinking) detonation.

Knocking (pinking)

(*a*) A periodic noise which is caused by parts or worn bearings.

(*b*) Refers to the noise from pre-ignition or detonation.

Knoop Hardness Test. Hardness test using an indenter in the form of
a four-sided pyramid whoose indentation has been a parallelogram with the longer diagonal about seven times that of the
shorter.

Knote (kn). A unit of speed; it is equal to one nautical mile per hour.

Knuckle Gearing. A gearing which is having teeth with a cross-sectional profile consisting of semi-circles above and below the pitch circle. This gearing has been strong and has been used for slow moving rough-purpose machinery.

Knuckle Joint. Refers to a hinged joint between two rods, a pin connecting the eye on one with a forked end on the other.

Knuckle Thread. A screw thread having semi-circular cross-section, and a radius one quarter of the pitch. This is a strong thread providing high friction, the latter necessitating generous clearance between male and female parts.

Knurling Tool (milling wheel). Small hard serrated steel rollers which are mounted on a pin, which get pressed against circular work to make a series of ridges on a surface to improve the finger grip on that surface.

Kollsman Altimeter. It is a sensitive altimeter in which bellows-type diagrams expand and contract with the chances in atmospheric pressure and control a rocking shaft whose moments get multiplied by gear-wheels to give altitude readings with three different indicators. A bimetal bracket is able to compensate the readings of the instrument for changes of temperature.

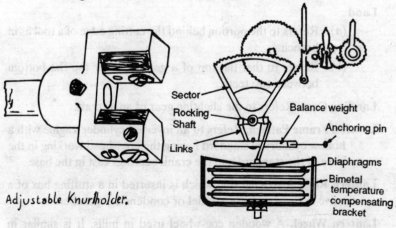

Adjustable Knurlholder.

K.B.B. Kollsman altimeter.

λ, **Lambda.** It is the symbol of wavelength.

Lb. Pound (force or mass).

Labyrinth, Gland Packing or Seal

    (*a*)   Refers to a series of grooves cut in the piston of a steam-engine to allow any escaping steam to expand slowly thus diminishing the leakage.

    (*b*)   A gland having radial and/or axial clearance.

Lace Machine. A machine in which bobbins, combs and carriages are able to convert two series of threads into an ornamental fabric.

    (*a*)   The delay in time between one event and the next.

    (*b*)   To provide thermal insulation.

Laminated. Made of thin plates or sheets called 'laminates'.

Land

    (*a*)   Refers to the portion behind the cutting edge of a tool as in branch.

    (*b*)   Refers to that flat top of a gear-tooth or the flat bottom between two teeth.

Landing Gear. Refers to the alighting gear of an aircraft.

Lantern Frame Pattern. Refers to an inverted cylinder engine with a hollow cylindrical standard having the crosshead working in the bore of the standard and the crank bearings cast in the base

Lantern Ring. A spacing ring which is inserted in a stuffing box of a valve to form a pressure relief or condensing chamber.

Lantern Wheel. A wooden cog-wheel used in mills. It is similar in design to a 'lantern pinion' and sometimes called a 'trundle wheel'.

Lap

    (*a*)   Refers to the amount by which one plate overlaps another.

(b) Refers to the amount by which a slide valve has to move from mid-position to open the steam or exhaust port of a steam-engine.

(c) Refers to the contact length of a chain around a sprocket wheel or of a belt around a pulley.

(d) Refers to thin sheet in which the fibre gets delivered to a carding engine after scutching and beating.

(e) A piece of soft metal, etc., or metal cylinder which is charged with polishing powder for lapping.

Lap Joint. Refers to a riveted or welded joint in which one member overlaps the other.

Lapping. Refers to the finishing and polishing of spindles, bearings, etc., to very fine limits by the use of laps of lead, brass, etc., machine.

Lapping Machine. A machine tool which is used for finishing the bore of cylinders, etc., using revolving laps and an abrasive powder suspended in the coolant.

Laser Marking. Refers to the marking of workpieces using a powerful laser to engrave the drawing number of serial number on to the finished part.

Lashlock. Refers to a split-and-sprung gear or nut in which relative motion of the two parts is prevented, on reversal of load, by a self-locking wedge having a light spring.

Lateral Traverse. Refers to the amount of end play given to locomotive trailing axles to allow the taking of sharp curves.

Lathe. A machine tool which consists of a lathe bed carrying a headstock and tailstack for driving and supporting the work, and of a saddle which carries the side rest for holding and, traversing the tool. It produces cylindrical work, boring facing and screw-cutting.

Lathe Bed. Refers to that part of the lathe which forms the support for the headstock, tailstock and carriage. It is a cast rigid box-section girder on legs having and the upper surface planned to provide a true working surface.

Lathe Carrier. A clamp which consists of a shank (either straight or bent) having an eye at one end and provided with a set-screw.

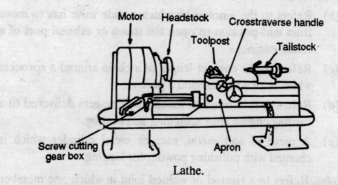

Lathe.

The clamp gets attached to work supported between centres and driven by the engagement of the driver plate pin with the shank of the carrier.

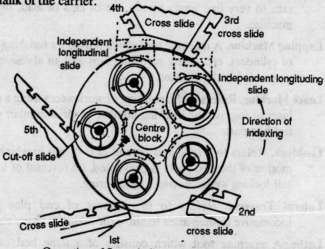

Operation of five-spindle automatic lathe.

Lathe Centre Grinder. An emery wheel with overhead drive for grinding the hardened conical points of lathe centres in place while mounted in the headstock or tailstock.

Lathe Cheeks. Refers to the sides of a lathe bed.

Lathe Heads. Refers to the headstock and tailstock of a lathe.

Lathe Planer. Refers to piece of mechanism, sometimes attached to the saddle of a lathe, for the surfacing of metal by rectilinear cutting, using a milling cutter in the head stock.

Lathe Standards. Refers to the supports of a lathe bed.

Lathe Tool. Any turning tool which is used in a lathe.

Lathe Traverse. Refers to the drive mechanism for moving the saddle to perform a turning or facing operation.

Lay Shaft. An auxiliary, or secondary, geared shaft.

Laying-out. Refers to the marking out or setting out of work, especially plate work, to full size ready for cutting, drilling etc.

Lazy Tongs. Refers to an arrangement of zig-zag levers for picking up objects.

Lead

    (*a*)   Refers to the distance between successive intersections of a helix by a generator of the cylinder on which it lies; the distance a screw thread advances in one revolution.

    (*b*)   Refers to the lead of the helix of which the tooth trace (in a gear) forms part.

Lead Angle

    (*a*)   The angle of a tooth trace for a helical, spur of worm gear.

    (*b*)   The acute angle between a tangent to the helix and a transverse plane. The complement of the helix angle.

Lead Line. The left-hand vertical line in an indicator diagram which is representing the rise in pressure at the start of the working stroke.

Lead of Valve. Refers to the amount by which the slide valve of a steam-engine has uncovered the port to steam when the position is at the beginning of its working stroke; sometimes called 'main screw'.

Lead Screw (leading screw, guide screw). The screw which is running longitudinally in front of the bed of a lathe and the master screw used for cutting a screw thread.

Leading Axle. Refers to the front axle of a locomotive.

Leading Edge. Refers to that edge of a wing, aerofoil, strut or propeller blade which first meets the air or water when the craft is in motion.

Leading Springs. Refers to the springs carrying the axle boxes of the leading wheel of locomotives and rolling work.

Leading Wheels. The front wheels of a locomotive.

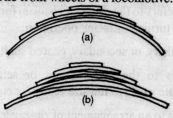

(a) Pulled up (b) Free
Leaf spring.

Leaf Spring (laminated spring). Refers to a curved (sometimes flat) spring having thin plates (leaves) superimposed and acting independently, to form a beam or cantilever of uniform strength. A simple flat plate spring is also known as a single leaf spring.

Least Angle of Traction. Friction angle.

Left-hand Engine. Refers to a horizontal engine which stands to the left of its flywheel as seen from the cylinder.

Left-hand Screw. It is a screw which turns anti-clockwise when being inserted.

Left-hand Thread. It is a screw thread which, when viewed along its axis, appears to rotate counter-clockwise as it goes away from the observer, the reverse of the common wood-screw.

Left-hand Tools. Lathe side tools having the cutting edge on the right, thereby cutting from left to right, that is away from the headstock of a conventional lathe.

Left-hand Twist Drill. A twist drill in which the cutting edge and flute are running anti-clockwise up the shank.

Left-handed Engine. As acro-engine in which the propeller shaft is rotating counter-clockwise when the observer has been looking past the engine to the propeller.

Lentz Valve-Gear. A valve-gear which is admitting and exhausting steam in a locomotive through two pairs of poppet valves, spring-controlled and operated from a camshaft rotating at engine speed.

Lever

    (*a*)   A rigid rod or beam which is pivoted at a point (fulcrum) with a load at one end and a force applied at the other.

    (*b*)   A pivoted arm which is carrying the pallet in a lever escapement.

Lever Box (lever bracket). A hollow casting in a crane which is carrying the levers connected to the motions for lifting, slewing, travelling etc.

Lever Chuck. A concentric chuck which is actuated by a lever instead of a screw.

Lever Jack

    (*a*)   A simple jack which is consisting of a lever for lifting and a standard for support.

    (*b*)   An accessory which is controlling the locker carriage. It is located on the underside of the combs of a lace machine. Also called 'locker jack'.

Lift (elevator)

    (*a*)   An enclosed platform which is working in a vertical shaft for transferring persons, goods or vehicles from one floor level to another and operating electrically hydralically or pneumatically with generally a winding drum or a traction sheave and a counterweight.

    (*b*)   Refers to the height to which a pump can raise water or other fluids.

Lift Valve. A valve in which the disc, ball, plate etc., lifts or is lifted, vertically, to allow the passage of a fluid.

Lifting Cylinder. A cylinder of a hydraulic crane which is used for lifting the load.

Lifting Jack. See jack.

Lifting-pump. See suction pump.

Lifting Ram. Refers to the smaller of the two rams in a hydraulic forging press which lifts the crosshead and tup after each stroke.

Lighting Cock. The jet that fires the charge in a gas-engine cylinder.

Light Running. Refers to the running of mechanisms, *e.g.*, shafting, under no load and with the minimum of friction.

Light-spring Diagram. An indicator diagram which is taken with a specially weak control spring or diaphragm to reproduce to a large scale the low-pressure part of the diagram.

Limit Gauge. Refers to a fixed gauge which is used for verifying that a part has been made within specified dimensional limits. It is either a 'go' or 'no-go' gauge.

Limit Gauging. It is a method of measurement to ensure the fitting of two pieces together within specified clearance limits and thus permitting inter-changeability.

Limit Load. Refers to the maximum load anticipated under normal conditions of operation of an aircraft.

Limit of Elasticity. See elastic limit.

Limit of Proportionality (creep limit). Refers to the point on the stress-strain curve at which the strain ceases to be proportional to the stress.

Limited Sequence Robot. Non-servo robot or pick-and-place device.

Limiting Friction. See friction.

Limiting Range of Stress. Refers to the greatest range of stress about a mean stress of zero that a metal can withstand for an indefinite number of cycles without failure. Also termed as 'endurance range': the fatigue (endurance) limit has been half this range.

Limits of Tolerance. See tolerance.

Lincoin Milling Machine. A milling machine having a fixed work-table height but vertically adjustable horizontally-mounted cutter shaft.

Line of Centres
  (*a*) A line which is passing through two or more centres in machinery or in a mechanism.
  (*b*) The line which is joining the balance staff and the pallet staff in a lever escapement.

Line Shafting. Refers to the main (overhead) shafting which is used in factors to transmit power from the power source to individual machines.

Line Standard. It is a standard of length. It is the distance between two fine lines on a metal bar which is measured under specified conditions. Line standards form the bases for checking gauges.

Linear Advance. Refers to the amount by which a slide valve is set forward for lap and lead beyond a line 90° ahead of the crank.

Linear Roller-bearing. It is a roller-bearing which allows linear motion, the rollers returning along a recirculation channel.

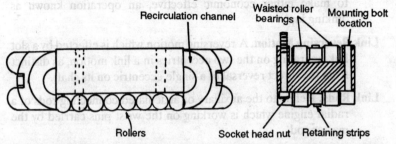

Linear roller-bearing.

Linear Sender. A device having a reciprocating action. It can be fitted to an electric drill, which is used for sandpapering or polishing depending upon the surface attached to the reciprocating part.

Linear Velocity. Velocity along a path which is either straight or curved.

Liner

    (*a*)   Refers to a separate and renewable sleeve kept within an engine cylinder to provide a more durable rubbing surface for the piston rings.

    (*b*)   Any sleeve which is fitted to provide a more durable surface.

Lining-up

    (*a*)   Refers to arranging the bearings of an engine crankshaft, etc., in perfect alignment.

    (*b*)   Refers to alignment of an assembly.

Link Belting. A belting (or belt) which is composed of a number of short links, arranged parallel and retained in position by pins, which allow the links to pivot freely and bend round small pulleys for transmitting power over a short distance.

Link Grinding Machine. A machine which is used for grinding the curves of the slot links of valve-gears, possessing spindles of planet type and the links moved about a centre, adjustable for radius.

Link Motion (stephenson's). A valve motion which is for reversing and controlling the cut-off of a steam-engine. It consists of a pair of eccentricts connected to the ends of a slotted link gets varied to make either economic effective, an operation known as 'linking lup'.

Link Reversing Motion. A reversing motion which is effected by a slot link operating on the two eccentrics in a link motion, as distinct from the direct reversal of a single eccentric on its shaft.

Link Rods. Refers to the auxiliary or articulated connecting rods of a radial engine which is working on the wrist pins carried by the master rod.

Linking. Refers to the possess in automation whereby articles in a transfer, line for machining or manufacture have been passed automatically, with inspection, between successive machines.

Lip. Refers to the cutting edge or point attached to a centre bit or similar tool, which cuts the circumscribing circle during the process of boring.

Lip Drill. A drill whose cutting faces have been slightly hollowed out backwards immediately above the cutting edges to give a front rake to the tool.

Lip Seal. It is a shaft seal which is used for retaining oil using an annular flexible sealing element with an interference fit on the shaft. The sealing pressure gets provided by the deformation stress and also a greater spring.

Liquid Spring. A person-cylinder combination which is used in aeroplane suspension units, the piston forcing hydralic fluid through a small hole.

Litre (l). It is a unit of volume which is equal to one cubic decimetre.

Live Axle. Driving axle.

Live Ring. A large roller-bearing for supporting turntables and revolving cranes.

Live Roller. A roller which is free to move, along its own path and rotate, but does not revolve on a spindle. Live rollers have been used for the siewing motions on heavy machinery and for turnable centres.

Live Steam. Steam supplied direct from a boiler.

Ljungstrom Turbine. It is a radial-flaw double-motion reaction turbine having groups of blades which are arranged in concentric rings. Alternative rings get attached to one of two discs mounted on separate contra-rotating shafts. The contra-rotation of the shafts effectively doubles the peripheral speed of all the rings except the first, thereby nearly doubling the steam velocity and giving an increase efficiency for a given overall size of turbine.

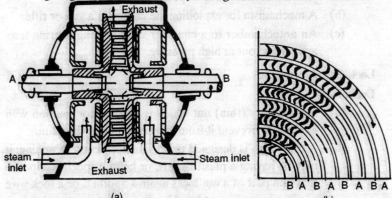

Ljungstrom turbine: (*a*) sectional view, (*b*) contra-rotation of shafts.

Load. Refers to the power output of an engine or powerplant under given circumstances.

Load-extension Curve. See stress-strain relationship.

Load Factor

(*a*)   Refers to the ratio of an average load to the maximum load.

(*b*)   Refers to the ratio of the external load on an aircraft in a specified flight condition to the weight of the aircraft.

(*c*)   Refers to the ratio of the number of passengers (or tons of frieght) in a vehicle to the maximum number (or load).

Loader. A mechanical shovel or similar device which is used for loading trucks.

Loading Diagram. It is a graphical representation of the loads and their locations acting on a body.

Lobe

(a)   Refers to a rounded projection or cam.

(b)   Refers to projections on an ignition contact-breaker.

(c)   The several cams which are formed on one ring in radial aero-engines.

(d)   The peripheral projections of a helical screw compressor.

Lock

(a)   A mechanical appliance which is used for fastening a door with a bolt that needs a special key to work it.

(b)   A mechanism for exploding the charge in a gun or rifle.

(c)   An antechamber to a chamber in which engineering tests are carried out at high pressure.

Lock. See steering lock.

Locknut

(a)   An auxiliary (thin) nut which is used in conjunction with another to prevent it from loosening under vibration.

(b)   A nut which is designed to obviate accidental loosening; it may be having a plastic insert, or be of a special shape, so that one part of a nut locks against another, or a lock wire or pin can get inserted and be appropriately named.

Lock Washer. A washer which is designed to prevent loosening of a nut of bolt head.

Locker Jack. Lever jack.

Locker Rack. A rack railway having the rack centrally located and with teeth on each side in which horizontal cog-wheels work.

Locking Face. Refers to the portion of a pallet upon which the teeth of the escape wheel drop for locking.

Locking Lever (locking piece). Refers to the lever that locks the chiming mechanism in a clock.

Locking Piece. Locking lever.

Locking Pin. The pin on the locking wheel.

Locking Wire. Wire which is used to secure nuts and pipe unions from vibrating loose.

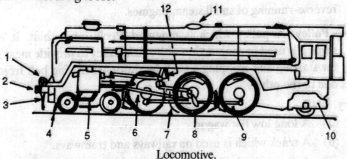

Locomotive.

Locomotive. A railway vehicle which is driven by steam, eletricity or oil for hauling trucks or carriages.

1. Vacuum brake connection.
2. Buffer.
3. Coupling link.
4. Guard iron.
5. Leading bogie.
6. Connecting rod.
7. Sand pipe.
8. Driving wheels.
9. Coupling rod.
10. Trailing truck.
11. Steam dome.
12. Valve gear.

Locomotive Boiler. A multi-tubular boiler.

Loom. Refers to a machine for weaving yarn or thread into fabric in which two sets of yarn or thread, 'warp' and 'weft', are interlaced.

Looming (healding). Drawing the threads of the warp through the eyes of the heald shaft, in the arranged order, prior to weaving, plus knotting and twisting.

Loose Coupling. A shaft coupling which is capable of instant disconnections.

Loose Eccentric. Refers to an eccentric riding freely on a shaft between two stops which position and drive it for ahead and reverse-running of small steam-engines.

Loose Pulley. A pulley which is mounted freely on a shaft. It is generally used in conjunction with a fast pulley to provide means for starting and stopping a shaft by shifting a driving belt from one to the other.

Lorry.

(a)   A long low flat wagon.

(b)   A truck which is used on railways and tramways.

(c)   A motor truck which is used for road transport.

Lost Motion. Refers to the difference between the rate of motion of driving and driven parts in a mechanism.

Low-pressure Cylinder. Refers to the largest cylinder of a multiple expansion steam-engine, in which the steam gets finally expanded.

Low-pressure Engine. Refers to an engine which exhausts its steam into a condenser.

Low-pressure Steam. Steam which is at a pressure below or only a little above atmospheric pressure.

Lubricant. A substance which is used for reducing friction between bearing surfaces in relative motion, such as oil, graphite, air under pressure etc.

Lubrication. Refers to the distribution of a lubricant between moving surfaces in contact to reduce the friction between them.

   Thick-film lubrication. Lubricant thick enough and limits of tolerance sufficient to disallow a metal-to-metal contact.

   Thin-film lubrication. Lubrication where a metal-to-metal contact exists and the characteristics of the mating surfaces significantly influence the bearing friction.

Lubricator. Any contrivance which is used for supplying a lubricant to bearing surfaces.

Luffing-jib Crane. A crane having its jib hinged at its lower end to the crane structure to allow alteration in its radius of action.

Lug. Far.

Lumen (Lm). May be refirmed as the luminous flux which is emitted within a unit solid angle of one steradian by a point source having a uniform intensity of one candela.

Lux (Lx). The unit of illumination = $1 \, 1m/m^2$.

M

μ Greek Letter Mu

    (a) Refers to the symbol for the coefficient of friction.

    (b) A micron equal to 10^{-3} mm.

    (c) μ in. A one-millionth of an inch.

M. Refers to the symbol for Moment.

M. Refers to the symbol for mass.

Mep. Abbreviation of Mean-effective-pressure.

Mech Eff. Abbreviation of mechanical efficiency.

MHN. Abbreviation of Meyer Hardness Number.

Mks Unit. A unit in the Meter-kilogramme-second System.

M of i., M.I. Abbreviation of Moment of Inertia.

MS. Abbreviation for maximum stress of Margin of Safety.

Mps. Metres per second. (SI notation: m/s).

Machine. An apparatus which consists of an assemblage of parts, some fixed and some movable, by which mechanical power get applied at one point, to transmit a force or motion at another point.

Machine Centres. Loose centres.

Machine Moulding. Refers to the process of making moulds and cores by mechanical means, replacing hand-ramming.

Machine Riveting. Clenching rivets by using hydraulic riveters or pneumatic riveters.

Machine Shop. It is a shop where all the operations of engineering requiring the use of machines are undertaken and excluding fitting and erecting.

Machine Tapper. A fitting which is attached to drilling machines or lathes for tapping threads in holes, with a spring release to prevent breakage on completion of the tapping operation followed by reversal of the driver for proving tap.

Machine Tools

(a) Tools which are used on various machines for cutting, drilling, milling, plaining, punching, shapping, shearing, slotting, turning etc.

(b) Machines which are used in manufacturing plant, including those referred to in (a).

Automatic machine tools. Machine tools having an unmanned repetitive action.

Machine Vice (vice chuck). Refers to a parallel-jawed vice which is used for holding pieces of work on the tables of drilling, planning and shapping machines.

Machining. Making or operating machines.

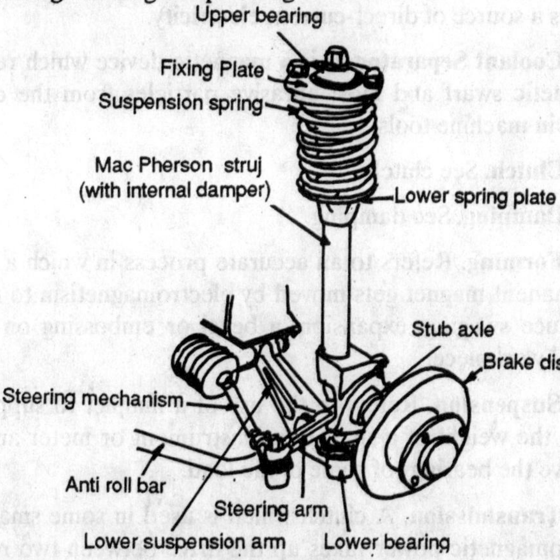

Fig. M.1 Mac pherson strut.

Machining Allowance. Material present on a casting or forging which is removed when producing a finished machined surface.

Mac Pherson Strut. The vertical support which is used with some motor vehicle front suspension systems. The upper end, attached to the vehicle body gives support for the top of the suspension spring which in turn gets located at its lower end by a plate attached to the movable lower half of the strut. Within the assembly has been a rubber bump stop and hydraulic damper. Which gets compressed when the spring and strut get compressed.

Magnetic Brake. A brake which is operated by means of an electromagnet.

Magnetic Chuck. Refers to a 'permanent magnet' chuck, which is having permanent magnets located in the base and separated from the top surface by moveable spacers having alternate rows of conductors, magnetic blocks and insulators, for holding light flat work securely on the table of a machine tool. The work-piece gets released by moving the spacers with a hand lever so that the insulators are aligned and prevent the magnetic flux from reaching the surface. An 'electromagnetic' chuck needs a source of direct-current electricity.

Magnetic Coolant Separator. It is a magnetic device which removes magnetic swarf and most abrasive particles from the coolant used in machine tools.

Magnetic Clutch. See clutch.

Magnetic Damping. See damping.

Magnetic Forming. Refers to an accurate process in which a coil or permanent magnet gets moved by electromagnetism to rapidly produce swaging, expansion, a bevel or embossing on a thin metal workpiece.

Magnetic Suspension. Refers to the use of a magnet to support, in part, the weight of a shaft in an instrument or meter and thus relieve the bearings of some of the load.

Magnetic Transmission. A clutch which is used in some small cars. Ferromagnetic power takes up the drive between two rotating members when it has been drawn into an annular gap by electromagnets.

Magnetostriction. Refers to the change in dimensions which are produced in certain magnetic materials when they are magnetized, as in iron, steel and especially nickel.

Main Bearings. The bearings for an engine's crankshaft.

Main Cylinder. Refers to the principle or working cylinder of an engine.

Main Driving Belt. Refers to the belt from the power unit to the main driving pulley in a workshop, whose machines have been belt-driven.

Main Driving Pulley. The principal pulley on a line of shafting.

Main Frames. Refers to the locomotive frames which carry the boiler, axle boxes, cylinder etc.

Main Rotor (S)

(a) Refers to the rotor(s) of a rotorcraft or hovercraft which provide lift, as distinct from the tail rotor.

(b) Refers to the assembly of compressors (s) and turbine (s) which comprise the rotating parts of a turbojet or turboprop engine.

Main Valve. It is the slide valve proper when there has been a separate expansion or cut-off valve for the stream.

Male and Female. Engineering terms which are applied to inner and outer members that fit together, such as threaded pieces, pipe fittings, etc.

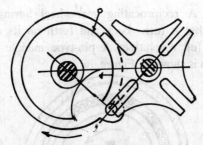

Maltese Cross mechanism (a) star wheel, (b) cam.

Mallets. Soft headed hammers which have been used to avoid damage to machined and other surfaces. Several types are used including the following: lead head with an iron handle or a wooden handle and iron head with an insert of copper, fibre hard rubber, box wood, plastic or hide.

Maltese Cross Mechanism (geneva wheel). Refers to mechanism which is used for feeding the film forward intermittently in a cinematograph projector (or cemera), involving a star-wheel and cam and obviating the use of claws. The mechanism also use in machine tools to give an intermittent action.

Manchester Principle. Diametral pitch.

Mandrel (mandril)

(*a*) A cylindrical rod which is usually parallel and sometimes tapered, upon which partly machined work gets mounted for turning, milling etc.

(*b*) Refers to a cylindrical rod used in the opening of a die to form the internal diameter when extruding tubes. When unsupported in the die it is termed a 'floating mandrel'.

(*c*) Refers to the driving or headstock spindle of a lathe:

Expanding mandrel. One which split and capable of expansion by a tapered plug.

Manifold Pressure (boost pressure). Refers to the absolute pressure in the induction manifold of an unsupercharged piston-engine.

When the engine gets supercharged, it is termed 'boost pressure'.

Mangle Wheel. A reciprocating gear-wheel having its teeth so arranged that it turns back and forth on its centre without making a full revolution. A pin-type mangle gear-reversing mechanism is shown in Figure.

Pin-type mangle gear-reversing mechanism; (a) guide pin, (b) pinion spindle.

Margin of Safety. The margin of safety, MS, may be defined as M/S = 1/R − 1. where R denotes the ratio of the applied load to the allowable load.

Marine Engine. Steam of compression-ignition (oil) engines which are used for ship propulsion and directly coupled to the propeller.

Marine Governor. A governor which is to control steam admission and thus check racing of the main shaft of a marine engine, when the propeller gets raised by the waves above the sea.

Marine Pattern Connecting Rod. A rod whose bearing end 'is having two brasses, secured by bolts to flat expansion of the wrought-iron end of the rod.

Marlborough Wheel. Refers to an extra wide gearwheel which makes it to mesh with more than one standard width gear wheel even if the standard width gears share a common radial disposition relative to the Marlborough wheel.

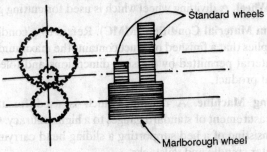

Marlborough wheel.

Marshall Valve-gear. Refers to a radial gear of Hackworth type in which the straight guide gets replaced by a curved slot to correct inequalities in steam distribution.

Masked Valve. A poppet valve having its head recessed into its seat so that its outer diameter acts as a piston valve thereby allowing a lower valve acceleration.

Mass Balance Weight

(a)   A mass which is added to a body produce a desired balance or inertia about some predetermined point.

(b)   A mass which is attached to an aircrafts control surface to reduce or eliminate the inertial coupling between the angular movement of the control and some other degree of freedom of the aircraft.

Mast. Refers to the vertical member in a derrick crane.

Master. A term which is used for a special gauge, tool, etc., or the key member of a system.

Master Connecting Rod (mother rod). A specially strengthened connecting rod which is used on one cylinder of an aero radial engine, which carries wrist pins to which the other connecting-rods gets articulated and thus transmits the total thrust of all cylinders to the crankpin.

Master Gear. A gear which is used as a reference standard.

Master Tap. An extremely accurate tap for use when great accuracy is desired.

Master Wheel. A dividing wheel which is used for cutting gear teeth.

Maximum Material Condition (MMC). Refers to a condition which implies that a finished product contains the maximum amount of material permitted by the size, dimensions and tolerances, for that product.

Measuring Machine. A machine which is used for the precise measurement of standard gauges to a high accuracy. One type is consisting of a bed supporting a sliding head carrying a micro-meter spindle and tailstocks.

Mechanical Admittance (mobility). The reciprocal of mechanical impedance.

Mechanical Advantage. Refers to the ratio of the load (or resistance) to the applied effort (or force) in a machine.

Mechanical Efficiency. Refers to the ratio of the brake horse-power of an engine to the indicated horse-power.

Mechanical Engineering. A branch of engineering which deals primarily with the design, production and operation of mechanisms and mechanical contrivances including prime-movers, vehicles and general engineering products.

Mechanical Equivalent of Heat. May be defined as the ratio of the mechanical energy being transformed into heat to the resulting quantity of heat generated. Its value is the joute (1J = 1 Nm) or 4.1885×10^7 ergs per calorie, or 775 fit-Ib/Btu.

Mechanical Impedance. Refers to the ratio of the total force acting in the direction of motion to the velocity at the point, or surface, of reference and for a specified frequency; for example, when a mechanical system has been vibrating with uniform amplitude and at the specified frequency. It has been the reciprocal or 'mechanical admittance.

Mechanical Refrigerator. Refers to a refrigerator, consisting of a compressor raising the pressure of the refrigerant; a condenser for removing the latent heat; a regulating office to lower the temperature and pressure; and an evaporator to absorb heat at a low temperature.

Mechanical Shovel. Refers to an excavating machine having a boom and a bucket lifting system with power for operation, such as by a diesel engine driving a pump or pumps to provide hydraulic power for the vehicle drive.

Mechanical Stoker (automatic stoker). A device which is used for supplying solid fuel continuously by gravity and in some cases carrying the fuel on an endless chain progressively through a furnace and depositing the ash.

Mechanics. May be defined as that branch of science and technology which studies the action of forces on bodies and of the motions they produce. 'States' has been the section which deals with forces in equalibrium; 'dynamics' has been the section concerned with the motions in relation to the forces; 'Kinematics' deals with the theory of the motion without reference to the forces. 'Kinetics' has been the science of the relations between motions of bodies and the forces acting on them.

Mechanization (mechanisation). Means the change from animal to mechanical power in transport and industry.

Mechanomotive Force. Refers to the root mean-square value of an alternating mechanical force. It is expressed in newtons or pounds force.

Megadyne. A unit of force equal to one million dynes (10N)

Mesh

    (*a*)   Refers to the state of gears when in contact.

    (*b*)   Refers to the size of the openings in gratings, sieves etc.

Meshed. A term implying that a gear, or system of gears, has been ready for power to be transmitted through the gear.

Metal Detector. An electrical instrument which is used for detecting stray metal parts in non-metallic raw-materials and finished parts. It usually gives visible and audible warnings, ejects the item or stops the production flow.

Metal Sawing Machine. A machine which is used for sawing metal bars, tubes etc., which are held in a machine vice while a reciprocating powered hacksaw cuts through them.

Metal Spinning. Refers to the shaping of sheet-metal disc into circular or moulded shapes on a lathe face-plate by the application of lateral pressure.

Metal Spraying (powder spraying). Refers to the application of a metal surface by spraying molten metal from a gun, possibly with ionization. The coating metal could be supplied to a gun as a powder or a thin rod.

Metal Spring Coupling. Spring steel loops which are set in axial slots in a pair of adjacent flanges attached to two independent shafts. The couplings allow transmission of torque between the shafts, even if the coaxial alignment is poor.

Metal Tolerance. See tolerance.

Metre (m). The metre may be defined the length equal to 1650 763.73 wavelengths to vacuum of the radiation corresponding to the transition between the levels $2p_{10}$ and $5d_5$ of the krypton 86 atom.

Meyer Hardness Number. It is a number which is obtained by the same test as for the Brinell hardness number; it refers to the ratio of the load divided by the projected are of the indentation.

M.G. Machine. Abbreviation of single-cylinder machine.

Michell Bearing. Refers to bearing in which pivoted pads support the thrust collar or journal and tilt slightly under the wedging action of the lubricant induced by their relative motion. This action provides improved lubrication conditions, a low friction coefficient and a low power loss in the bearing.

Pivot segment bearing. The bearings are used for marine propeller shafts and other heavy-duty machinery.

Micro-drilling. Refers to the drilling of minute holes using very small drills; for example, one of 5 micrometres (0.002 in) diameter.

Microinch (μin). It is a unit for designating surface roughness which is equal to one millionth of an inch.

Micrometer. An instrument with optical magnification which is used. for measuring visually small angular separations.

Micrometer Gauge

(*a*)   A length gauge which is using two smooth faces connected by a horse-shoe, the gap between the measuring face being adjustable by an accurate screw at one face. The gap has been read off from a circular scale engraved under the timble head of the screw.

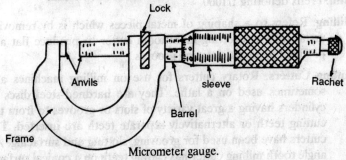

Micrometer gauge.

(*b*)   *Inside micrometer gauge.* These cylindrical instruments are having a fixed anvil at one end a moving anvil at the other. They could be extended in length using extension rods of fixed length in place of the nominal length fixed anvil. The barrel and timble assembly, the micrometer head common to all micrometer, can get attached to any measuring instrument.

Mid (or middle) Gear. Refers to the position of a steam-engine link motion or valve-gear when the valve motion has been a minimum.

Mil. Refers to the thousandth part of an inch. Colloquially called a 'thou'.

Mill

   (*a*) A machine, or building which is fitted with machinery for manufacturing processes.

   (*b*) A machine which is used for grinding, crushing or rolling.

   (*c*) A grinding mill where the millstone is running round on a horizontal arbor and about a central vertical shaft.

Mill Engine. Refers to a large, low-speed horizontal stem-engine which is fitted with drop valves or Corliss valves or a Unaflow engine. It is sometimes used to drive machinery through ropes.

Mill Gearing. Gearing which is comprising cog-wheels, pulleys, shaft bearings and belting.

Milled Head (knurled head). Refers to the head of an adjusting screw roughened or cut in a succession of ridges to provide a good grip.

Milli. Prefix denoting 1/1000.

Milling. Refers to a shaping of metal pieces which is by removing metal with a revolving multi-tooth cutter to produce flat and profilled surfaces, slots and grooves.

Milling Cutters. Rotary cutters for use on milling machines and sometimes used on a lathe. They are hardned-steel discs or cylinders having a great variety of slots or grooves to from the cutting teeth or alternatively separate teeth are inserted. The cutters have been used for grooving, slotting and surfacing. An angle tooth milling cutter has cutting teeth on a conical surface, typically at 45° to its axis, so that inclined surfaces can get machined.

Milling Machine

   (*a*) A machine tool having a horizontal arbor or a vertical spindle to carry a rotating multi-tooth cutter with the work supported and fed by an adjustable and power-driven table.

   (*b*) A rotary machine which consists of squeezing-rollers and a box channel called the spout, etc., fitted over a large trough, the whole being enclosed. It is used for the

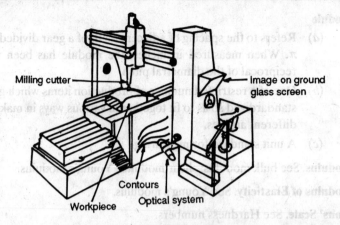

Counter milling machine.

preparation of woollen fabrics for a subsequent finishing process.

Milling Wheel. Knulrling tool.

Minus Lap. Refers to the exhaust lead on a steam valve for diminishing the amount of cushioning.

Misfiring. Refers to the failure of the mixture in the cylinder of an internal-combustion engine to fire normally, due to ignition failure or to an over-rich or too weak a mixture.

Mitre Gear. Refers to a pair of bevel gear-wheels in mesh having their shafts at right angles.

Mitre Valve. A safety valve having the annular seating cut at an angle of 45^0.

Mitre-cut Piston-ring. A piston-ring which is having the ends mitred at the joint; as distinct from stepped or square ends.

Mixed Flow (American) Water Turbine. A inward-flow reaction turbine having the curved runner vanes acted on by the water as it enters radially and leaves axially.

Mixed-pressure Turbine. A steam turbine which is operated from two or more sources of steam at different pressures.

Mixing Chamber. A chamber where fuel and air get mixed prior to ignition, especially in a gas-chamber.

Module

(a) Refers to the spacing of adjacent teeth of a gear divided by
     π. When measured in inches, the module has been the
     reciprocal of the diametral pitch.

(b) One of a restricted number of production items which gets
     standardized so as to fit together in various ways in making
     different articles.

(c) A unit standard for measuring.

Modulus. See bulk modulus; shear modulus; Young's modulus.

Modulus of Elasticity. See Young's modulus.

Mohs' Scale. See Hardness numbers.

Moment of Force. Refers to the turning effect of a force about a given
point which is measured by the product of the force and the
perpendicular distance of the point from the line of action of the
force. Generally, clockwise moments have been called 'positive'
and counter-clockwise moments have been called 'negative'
moments. Units are force times distance, *e.g.*, newton metres
(Nm), pounds force feet (1 bf ft.) or pounds-force inches.

Moment of Inertia. The sum Σmr^2, where m denotes the mass or a
particle in the body and r its perpendicular distance from the
axis.

Moment of Momentum (angular momentum). See momentum.

Momentum. Refers to the product of the mass of a body and its
velocity.

   Angular momentum. Refers to the product of the moment of
   inertia and the angular velocity of a body. The sum of all the
   momenta remains unaltered in any one mechanical system.

Monitor. An instrument which is used for keeping a variable quantity
within definite limits by transmitting a controlling signals, as in a
process plant.

Monkey. The falling weight which is used in a pile driver.

Monkey Wheel. Gin block.

Monobloc. Refers to the integral casting of all the cylinders of an
internal combustion engine in one block.

Monocable. Refers to an aerial railway in which a single endless rope both supports and moves the loads.

Morse Tapers. Standard tapers which are for fitting the shanks of drills, etc., to machine spindles.

Mortise Teeth. Cogs.

Mortising Machine. A machine which is used for cutting square or rectangular holes in wood. The reciprocating solid chisel and rotary bit in the older type machines are replaced by the hollow chisel or the edged chain or both in combination in modern machines. The hollow chisel has been especially particularly suitable for the automatic mortising machine.

Motion Bars. Guide bars.

Motion Block. Refers to a block, attached to the valve rod of some steam-engine valve-gears, which gets constrained to move in a circular path by a curved slotted link.

Motion Disc. Wrist plate.

Motor.
  (*a*)   Refers to a prime mover.
  (*b*)   Refers to the petrol engine of a motor-car or aircraft.

Motor Car (automobile). A private motor-vehicle.

Motor Vehicle (automobile). A road vehicle which is powered by a petrol engine or diesel engine.

Mould
  (*a*)   A hollow receptacle of wood or metal having a special sand for casting into which the pattern is pressed.
  (*b*)   The permanent metal shape which is used for die-casting.

  Moulding box. A box divided into sections so that it can be taken apart for removal of a complex casting.

Moulding Cutter. An adjustable and specially shaped revolving cutter. It is often used in pairs on opposite sides, for cutting a desired moulding profile.

Moulding Machine (plastics). Machines which are used for moulding plastics by compression and transfer mouldings are shown in Figure. The plastic may be taken in powder or granular form or

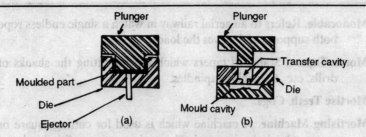

Moulding machines (a) compression, (b) transfer.

in pellets of plastic powder (performs) in the proper amount for
making the moulded part. Compressior moulding uses
pressures of 14MPa to 140MPa (2000 to 20000 Ibf/in^2) and a
mould heated to soften the plastic. In transfer moulding, the
dies have been closed before the plastic has been added in the
cylinder above. Injection moulding (Fig. 21) has been similar to
transfer moulding except that the soft plastic has been forced
into the die cavity under pressure which has been maintained
until the plastic has been cooled by water circulating in the walls

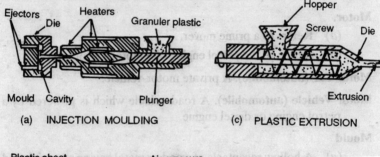

(a) INJECTION MOULDING   (c) PLASTIC EXTRUSION

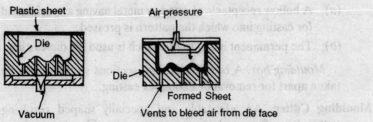

(b) BLOWING AND VACUUM FORMING

Plastics moulding.

of the die. Extrusion, blowing and laminating processes have been also used for plastics. Fig. 2 shows the process of plastic extrusion, in which the granulated powder gets fed into a hopper, then passes into a conveyor screw where the heat is applied to soften the plastic sufficiently to flow through the die. Blowing and vacuum-forming a single die on a plastic sheet.

Mounting. Refers to the chucking of work in a lathe.

Mouse Roller. A small extra roller which is used to improve the distribution of ink in a printing machine.

Movable Expansion. Expansion which is capable of regulation by means of a second slide valve or other gear in a steam engine.

Moving-iron Instrument. An instrument which is depending on the movement of a piece of moving soft iron relative to a magnet as shown typically in the air dashpot which has been an 'attraction' type of instrument. The movement between two mutually repulsive magnetized piece of iron, the one fixed and the other controlled by a spring is termed as the 'repulsive' type of instrument.

Moving Staircase. Escalator.

M Teeth. Saw-teeth, shaped like the letter M, there are used in some cross-cut saws.

Mud Bucket. Refers to the bucket or scoop of a dredger.

Muff Coupling. Box coupling.

Multi-plate Disc Brake. A disc brake having several system acting in unison on the same rotating shaft.

Multiple Boring Machine. A boring machine having several mandrels for simultaneous boring.

Multiple-expansion Engine. Refers to an engine in which the expansion of the working fluid has been in two or more stages through cylinders of increasing size.

Multiple-spindle Drilling Machine. A drilling machine with several vertical spindles for simultaneous operation on a piece of work.

Multiple-threaded Screw (multi-start thread). A screw of course pitch having several threads to reduce the size of thread, to increase the relative size of the core and to obtain a higher velocity ratio.

Multiple-tool Lathe. A heavy lathe having two large tool-posts, one on each side of the work, carrying separate tools to operate simultaneously on different parts of the work.

Multiplier Register. Refers to the register in a calculating machine which records the number of turns of the multiplying handle (or its equivalent).

Multi-stage Pump. A centrifugal pump having two or more impellers mounted on the same shaft.

Multi-start Thread. Multi-threaded screw.

Mushroom Valve. Poppet valve.

N

N. Refers to the symbol for a newton.

V. Greek letter Nu. The symbol for kinematic viscosity $= \mu p$: viscosity $\div$ density.

N. Refers to the symbol of revolutions per unit-time.

NA. Refers to Neutral Axis.

NHp. Nominal Horse-power.

NTP. Normal Temperature and Pressure, *i.e.* 0°C and 760 mm of mercury.

Narrow Gauge. A railway gauge which is less than the 4ft 8½ in.

Nave. Refers to the hub of a wheel.

Necking. This term means the decrease in cross-sectional area over a short longitudinal length just prior to tensile failure. It can generally be observed in test specimens undergoing a tensile test.

Needle Lubricator. An inverted stoppered flask which is fitted to a bearing with a wire loosely fitting in the stopper and touching the shaft.

Needle Valve

    (*a*) A valve which consists of a thin tapered rod (needle) passing through a circular hole in a plate (orifice) or the end of a pipe and restricting the flow of a fluid. Relatively crude axial movement of the needle causes a small but regular change in the cross-section through which the fluid can flow. These valves are generally found in carburetters.

    (*b*) A screw-down stop valve which may be having the body ends in line or at right-angles to each other or may be of the oblique type with the disc in the form of a needle point. Needle valves are usually restricted to small sizes.

Negative Lead. The term used for the amount by which a steam port gets closed to admission when the piston has been at the bottom of the cylinder.

Netting Machine (netloom). A machine which produces netting with the threads knotted at their intersections.

Neutral Axis. Refers to the line through the locations of zero stress, of a number of cross-sections.

Neutral Gear. When the gearing of a car has been so arranged that no power can get transmitted, the engine or car is said to be in neutral gear.

Neutral Surface. Refers to the longitudinal surface of zero stress, therefore having the neutral axis, in a member which is subjected to bending.

Newton (N). May be defined as the force required to accelerate a mass of one kilogram at one metre per second squared.

Newton's Laws. These three laws were first stated by Newton in his 'Principia' and form the basis of all classical mechanics.

(*i*)   Every body remains at rest, moving at a constant speed in a straight line, unless it is acted upon by an external force.

(*ii*) If an external force acts on a body it will accelerate it in proportion to the size of the force and inversely in proportion to the mass of the body.

(*iii*) Every action is opposed by an equal and opposite reaction.

Nipple. A small drilled bush or tubular nut, or a short length of externally threaded pipe.

Nadal Gearing. Refers to the location of gear-wheels at a nodal point of a shaft system.

Nominal Horsepower. It is an obsolete method of rating steam-engines; for a piston-engine it if $D^2N(S)^{1/3}/15.6$, where S denotes the stroke in feet. N denotes the number of cylinders and D denotes the cylinder diameter in inches.

Non-carrying Conveyors. Conveyors which are chiefly of granular products. In these the material gets pushed along a fixed duct by blades mounted on a moving chain, belt or conveyor screw.

Non-condensing Engine. A type engine which exhausts its steam direct into the atmosphere.

Non-destructive Testing. Refers to any form of testing which in used to verify the integrity and expected full service life of an item which does not itself reduce the longevity of that item.

Non-servo Robot. It is a robot without feedback concerning the position of its elements relative to their nominal desired positions. Also called limited sequence robot, bang-bang robot or pick-and-place device.

Normal Pitch. Means the pitch of the traces of adjacent corresponding tooth flanks of helical, spur and worm gears which is measured along a common normal.

Normal Pitch (parallel helices). Refers to the distance between adjacent intersections of a system of concentric helices with a co-cylindrical normal helix measured along the latter.

Normally Aspirated Engine. Refers to a petrol or oil engine without supercharge or boost.

Nose. The term used for the front part of a spindle, mandrel or some projecting part.

Nose Cap. A boss or nub fairing fitted coaxially and rotating with a propeller, but not extending beyond the blade roots.

Notch. The term used for geometric discontinuity in a material which has been visible using pure optical methods. This type of discontinuity is of macro proportions $10 \mu m$ (3.94×10^{-4} in) to 40 μm (1.57×10^{-3} in.).

Notch Brittleness. Refers to the brittle property of a material causing fracture with small absorption of energy in an Izod or Charpy test.

Notch-sensitivity Ratio. This ratio has been found to vary from zero, for some soft ductile materials, to one, for certain hard and brittle materials. It may be defined as $(k_f - 1)/(k_t - 1)$ where k_f denotes the fatigue-strength reduction factor (see stress concentration factor) and k_t denotes the stress concentration factor.

Noteched-bar Test (impact test, izod test). Subjecting a notched metal test-piece to a sudden blow by a striker which is performed by a pendulum or falling weight by which the energy of fracture is measured.

Notching. A process which is used to cut a configuration of indentations in the edges of sheet-metal parts.

Nozzle

(a) Refers to an outlet tube through which a fluid escapes from a container.

(b) Refers to a specially shaped passage for expanding in impulse turbines.

(c) Refers to an injection fuel valve for oil- engines.

Nozzle Guide Valve. The term used for a ring of radially-positioned vanes, shaped like aerofoils which are able accelerate the gases from the combustion chamber of a gas-turbine type of engine and direct them on to the first rotating turbine stage.

Numerical Control. The term used for the operation of machine tools from numerical data previously stored on magnetic discs or tapes, punched tapes or punched cards. The machine data is generally obtained by computer from the design data.

Nusselt Number. A coefficient of heat transfer which may be defined by the ratio

$$Nu = Ql/kS\Delta T = hl/k$$

where Q denotes the quantity of heat transferred per unit time from an area S, K denotes the thermal conductivity; ΔT the temperature difference between two representative points; 1 a representative length and h is the heat transfer coefficient.

Nut. Refers to the mating part of screwed members which get rotated to tighten their hold. The heads may have various shapes, square, hexagonal etc., to fit the spanners used to rotate them.

Nut Runner. A power tool, fitted with a socket for an appropriate sized nut, used to tighten nuts, usually to a predetermined torque. Frequently it can also be reversed to loosen nuts.

Nutation. The periodic variation of the direction of a spin axis. Generally, the spin axis mutates in the form of a circular cone.

ω. Refers to the symbol for angular velocity.

O-ring (O-seal). Refers to a toroidal ring of circular cross-section which is made of rubber, neoprene or similar material usually fitting into a carefully machioned groove to provide a sealing between two making parts.

Oblique Projection. Refers to a projection in which two or three mutually perpendicular edges have been drawn at right-angles while the third has been at any angles to the horizontal, usually 30° or 45°.(See Figure 1).

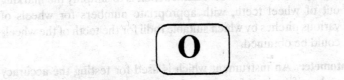

Fig. 1. Oblique projection.

Oblique Valve. It is a screw-down stop-valve having casing or body of spherical shape. The axis of the stem has been oblique to the body ends which are in line with each other.

Obliquity of Connecting-rod. Refers to the angle made by the connecting rod with the cylinder axis of a steam-engine when the crank pin has been at the extreme upper and lower portions of its path.

Obturator Ring. A gas ring L-shaped in cross-section which is present on the piston of a piston-engine to maintain a gas-tight seal between the piston and the cylinder wall.

Octane Number. Refers to the number indicating the knock rating of a motor fuel.

Odometer. An instrumented sheave which is used for recording the amount of line paid out, as when making oceanographic depth soundings.

Odontograph. A scale whose main function is to simplify the marking out of wheel teeth, with appropriate numbers for wheels of various pitches by which suitable radii for the teeth of the wheels could be obtained.

Odontometer. An instrument which is used for testing the accuracy and uniformity of gear-tooth profiles and tooth spacings in production work.

Off-line Programming. Programming of a machine or a robot which is using a programming language.

Ohm (Ω). May be defined as the resistance between two points on a conductor at a potential difference of one volt when a current of one ampere has been flowing.

Oil Cataract. Oil cylinder.

Oil Cooled. An engine or mechanism which have been immersed in oil to facilitate cooling.

Oil Cylinder (oil cataract). A small cylinder which is able to control the amount of piston movement in a steam reversing cylinder, the oil pressure being regulated by a cock.

Oil Engine. Compression-ignition engine.

Oil Feed

    (*a*)   Any appliance which is feeding oil to a bearing or some moving part of an engine or mechanism.

    (*b*)   The system of pipes and manifolds which are used to supply diesel oil (fuel) to the fuel injector on each cylinder of a compression-ignition engine.

Oil Grooves. Grooves which have been cut in the sliding faces, bearing surfaces, etc., for the distribution of lubricating oil.

Oil Hardening. The term used for the hardening of cutting tools of high carbon content by heating and then quenching in oil, which cools it less suddenly than in water.

Oil Pump. It is a small auxiliary pump which is driven from an internal-combustion engine crankshaft to force oil from sump oil tank to the bearings.

Oil Ring. Scraper ring.

Oil Sump. Refers to the lower part of the crankcase of an internal combustion engine which is acting as an oil reservoir.

Oiling Ring. A light metal ring which is riding loosely on a shaft located in a slot of the upper brass of a journal bearing. As the ring rotates it feeds oil to the brasses from an oil reservoir in the base of the housing into which the ring dips.

Oils. Neutral liquids which are used for lubrication with three main classes:

(1) Fixed (fatty) oils from animal, vegetable and marine sources, mainly glycerides and esters of fatty acids.

(2) Minerals oils from petroleum, coal, etc., which have been hydrocarbons.

(3) Essential oils, in the form of volatile products from certain plants, which have been hydrocarbons.

Oil-sealing Ring. A ring outside a roller-bearing which is used to prevent the escape of oil from the bearing.

Oilways. Holes drilled in a shaft or crankshaft to supply oil to the bearings supporting it, via oil grooves.

Oldham Coupling (double slider coupling). A pair to flanges, having opposed faces carrying diametral slots, between which a floating disc has been supported through corresponding diametral tongues arranged at right-angles so as to connect two misaligned shafts. (Figure 2).

Oilgocyclic Stress. Refers to the level of stress in a body which, although insufficient to cause immediate failure, would bring about failure after only a few repeated applications has been termed as the oilgocyclic stress. The term has been used to define the lower point of interest on an S-N curve.

Fig. 2. Oldham coupling.

Oliver

(*a*) Refers to a simple form of power hammer which finds use in some branches of chain-making.

(*b*) Refers to a small-lift hammer used by smiths. It is consisting of a horizontal shaft on end bearings with a hammer at the end of the shaft operated by a treadle underneath and a spring pole overhead.

Omtimeter (optimeter). It is an optical projection measuring instrument of high precision with a comparison measuring scale magnified about 100 times. It is used for comparing screw threads against a standard.

Opening Die (self-opening die). A die which clears the screw thread when it comes to the end.

Opposed-cylinder Engine. It is an internal-combustion engine having cylinders on opposite sides of the crankcase and in the same plane, having their connecting-rods working on a common crankshaft placed between them.

Opposed-piston Engine. It is an engine having a pair of pistons in a common cylinder the explosive mixture being ignited between the two pistons. One type of opposed-piston two-stroke compression-ignition engine is having triangulated form for the cylinder blocks, with three crankshafts at the apices of the equilateral triangle.

Optical Flat. Refers to a surface which generally of glass or quartz, and is having deviations from a truly plane surface which are small in comparison with the wavelength of light. It finds use to measure errors in the flatness of precision-finished surfaces by the interference fringes which can be seen when the optical flat is laid on the other surface.

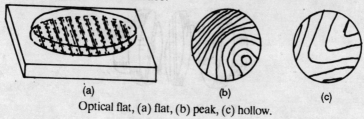

Optical flat, (a) flat, (b) peak, (c) hollow.

Optical Indicator. An engine indicator which uses optical methods to project an indicator diagram on a glass screen or for recording on a photographic plate.

Optical Tooling. An optical method which is used for checking the alignment of bearings, frames etc.

Optimeter. Optimeter.

Order Number. May be defined as the number of vibrations or impulses which occur per revolution during the torsional oscillations of an engine crankshaft.

Orthotropic Material. Refers to a material whose elastic properties vary in different planes, *e.g.*, timber and some reinforced plastics.

Oscillation, Centre of. Centre of oscillation.

Otto Cycle. It is a working cycle of a four-stroke piston-engine involving suction, compression, explosion at constant volume, expansion and exhaust, it involves reversible heating and reversible cooling at constant volume.

Out of Gear. When the wheels of a gear train have been disengaged.

Out Stroke. Refers to the stroke of a gas-engine piston in a direction away from the ignition chamber.

Outer Dead Centre (bottom dead centre). Refers to the piston position in a piston-engine or pump nearest the crankshaft when the piston has been at the end of its out stroke.

Out-of-balance. A rotating part has been said to be out-of-balance if rotation generates a resultant non-axial force.

Outside Crank. A single-web crank which is attached to a crankshaft outside the main bearings.

Outside Cylinders. Locomotive cylinders which are carried outside the frame and are working on to crank pins in the driving wheels.

Outside Lap (steamlap). Refers to the overlap of the slide valve of a steam-engine beyond the edge of the steam ports when in mid-position.

Outside Screw Tools. Chasers.

Outward Flow Turbine. See turbine.

Oval Chuck. Refers to a compound chuck in which the eccentricity gets controlled by a worm wheel and a tangent screw.

Oval Hole Cutting. Refers to the hole which is made by a cutter controlled in a similar manner to that of an oval chuck.

Overdrive. Refers to a device which is used for reducing the gear ratio in a motor-vehicle under optimum driving conditions to provide greater speed and to decrease fuel consumption.

Over-grip Sensor. A sensor which is able to know the presence of a component to be determined. If the end effector closes beyond its normal position, the component is assumed to be missing.

Overhanging Cylinders. Engine cylinders which have been bolted to the ends of their bed plates instead of upon their faces. This lowers the piston-rod centre and shortens the foundation for the bed.

Overhanging Shaft. Refers to the end portion of a shaft which is overhanging beyond its last bearing.

Overhead Camshaft. Refers to a camshaft which is running across the top of the cylinder heads of an engine and usually driven by a bevel-shaft or timing chain from the crankshaft. The cams operate on rockers or directly on valve-stems.

Overhead Gear. Machinery working overhead.

Overhead Valves. Inlet and exhaust valves which are working in the cylinder head opposite the piston in a vertical petrol or oil engine.

Overlap Ratio

    (*a*)   Refers to the ratio of the face-width to the axial pitch of the helical gear.

    (*b*)   Refers to the ratio of the angle subtended at the apex of the developed pitch cone of a bevel gear by the tooth trace, θ, to the angle subtended at the same apex by two points on the pitch circle and the similar flanks of adjacent teeth, ϕ.

Overload Coupling. Refers to a coupling which specially designed so that when a preset torque is exceeded the transmission of power is terminated.

Overriding. See riding.

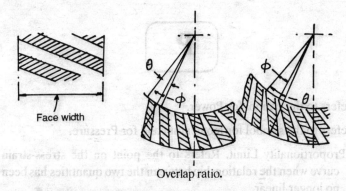

Face width

Overlap ratio.

Overstrain. Refers to the result of stressing an elastic material beyond its yield point.

Oxyacetylene Cutting. The same process as welding, the cut being started by a jet of oxygen on to the metal after preheating.

Oxyacetylene Welding. See welding; gas welding.

Oxygen Lancing. Refers to the cutting of heavy sections of cast iron or steel with oxygen fed to the cutting region via a steel tube which has been used up as the cutting proceeds.

P

P. Refers to the symbol for Power.

P. Refers to the symbol for momentum and for Pressure.

PL. Proportionality Limit. Refers to the point on the stress-strain curve when the relationship between the two quantities has been no longer linear.

PS. Proof Stress.

Packing. Material which has been inserted in stuffing boxes to make engine and pump rods pressure-tight.

> *Hydraulic packing.* Rings of 'L' or 'V' cross-section that will provide a self-tightening packing under fluid pressure.

> *Metallic packing.* A packing which consists of a number of soft metal rings or helix of metallic yarn, which encircles a piston rod and gets pressed into contact with the rod by a gland nut.

Paddle Shaft. Refers to the paddle-wheel shaft which has been driven directly by the engine cranks.

Paddle Wheel. Wheels at the sides or stern of a ship which are fitted with blades parallel to the shaft that dip into the water to propel the vessel. The blades (float blades) can be fixed or feathered.

Paddle Wheel Fan. Centrifugal fan.

Paddle Wheel Hopper. Refers to a feed hopper in which a rotating paddle wheel supplies small objects to a delivery chute.

Pallet
- (*a*)   A stand or container which is adapted for transportation of goods by fork-lift truck.
- (*b*)   A platform which is used to facilitate the handling of stacked items specially by a fork-lift truck, the pallet having suitable holes to accept the prongs of the fork.

Panel Beating. The term used for the manual or machine process by which sheet metal takes up a complex shape after stretch forming, gathering, drawing etc.

Pantograph. A mechanism which is based on the geometry of a parallelogram, for copying plans, etc., to a different scale.

Parabolic Governor. Crossed-arm governor.

Parallel Gate Valve. Double disc (gate) valve.

Parallel Motion

(a) A system of links which copying a reciprocating motion to an enlarged scale, such as on piston-type engine indicators.

(b) A system of levers which are adapting a curved movement to a rectilinear reciprocating motion.

Parallel Screw Thread. A screw thread which has been cut on the surface of a cylinder.

Parallel Slide Valve. Refers to a gate valve having one or two discs sliding between parallel body seats without a spreading mechanism as in a double disc (gate) valve. The effective closure would be obtained by the pressure of the fluid forcing the downstream disc-face against its mating body seat.

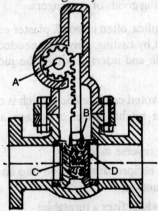

Parallel slide valve, (a) actuating spindle, (b) carrier, (c) spring for separating discs, (d) discs.

Paris (wire) Gauge. A metric gauge having numbers for wire diameters from no. 1 (0.6 mm) to no. 30 (10 mm).

Parson's Steam Turbine. Refers to a reaction turbine in which rings of moving blades of increasing size have been arranged along the periphery of a drum of increasing diameter. Fixed blades in the casing alternate with these rings. Steam expands gradually through the blading, from inlet pressure at the smallest section to condenser pressure at the other end.

Parting-off Tool. A tool which is used for metal and wood turning. It is narrow, deep, square across the end, and the width tapering slightly backwards so that it will clear itself in the cut and remove the material from the workpiece held in the chuck.

Pascal (Pa). A unit of pressure $= 1 \text{ N/m}^2$.

Passage. Refers to the steam-ways of a cylinder including both ports and exhaust.

Passes. Refers to the passing and repassing of bars, etc., through the rolls of a rolling mill. A backward pass over the top of a two-high mill is termed as 'lost pass'.

Passing Hollow. Crescent.

Paternoster. Refers to a series of floor sections which are moving slowly on an endless belt or chain, following gradual undulations if needed carrying goods or passengers.

Pattern. Refers to replica, often in wood, plaster or plastic, of the part to be produced by casting. Any hole needed in the casting are previously made and inserted within the mould before casting takes place.

Pawl (paul). It is a pivoted catch or click which is engaged by an edge or hook with a ratchet wheel or a rack and usually spring-controlled
- (*a*) to prevent reverse motion; or
- (*b*) to convert reciprocating motion into an intermittent rotary or linear motion;
- (*c*) the catch which fixes a turntable;
- (*d*) the catch which disallows the winch shaft of a crane from sliding when the gears are changed.

Pawl Feed (ratchet feed). Refers to the feed of a machine which is effected by means of a pawl and ratchet or small cogwheel.

Peaucellier Mechanism. A four-bar rhomboid mechanism, ABCD, have been constrained by equal-length links OB and OD to a point, O, on the end of a diameter of a circle and the point C moves on the circumference of the circle. This mechanism has been constrained the point A to move on a line perpendicular to the diameter on the end of which is the point O. This mechanism produces a straight-line motion.

Pedestal. A support for a shaft or other part of a mechanism.

Pedestal Bearing. Refers to a bearing which is held in a cast, forged or welded support which is frequently bolted to a main structure. Sometimes it has been split at the centreline forming a plummer block.

Pedometer. An instrument which is used for recording the number of steps taken by a pedestrian, actuated at each step by the movement of a small weight which is balanced against spring.

Pelton Wheel. Refers to an impulse water turbine which is rotated by a jet of water from a nozzle striking specially shaped buckets attached to the periphery of the wheel, the nozzle being either deflected or valve-controlled by a governor.

Pendulum. A body which has been suspended so as to be free to swing, especially a rod with a weight at the end for regulating the movements of clock's works. Household clocks beat half seconds or less; longcase clocks and regulators have a seconds pendulum; tower clocks have pendulums which may beat up to two seconds.

Pendulum Bob. Refers to the weight at the bottom end of a pendulum.

Pendulum Damper. Pivoted balance weights which have been attached to the crank of a radial piston-engine for neutralizing the fundamental torque impulses and thereby eliminating the associated critical speed.

Pendulum Governor. Refers to an engine governor in which heavy balls swing outwards under centrifugal force, thus lifting a weighted sleeve and progressively closing the engine throttle valve.

Pendulum Rolling Mill. Refers to a rolling mill which is used to achieve very large reductions, up to 50% at a single pass, in the thickness of sheet-metals which are otherwise difficult to fabricate.

Perambulator. A surveying instrument which used for distance measurement. It consists of a large wheel supported on its axis by a long handle to wheel it along the distance to be measured, with a revolutions recorder.

Percusssion, Centre of. Centre of oscillation.

Peripheral Speed. Refers to the speed of any point on the periphery of a rotating wheel, cutter etc.

Peristaltic Pump. A pump which produces a flow in a fluid within a flexible pipe by employing a series of roller each of which in turn is passing over the same given length of pipe. Each roller squeezes the tube progressively, forcing the fluid along the given length but before a roller releases its hold that next one starts its traverse hence the fluid has been continuously pumped along the pipe.

With this type of low-pressure pump the fluid remains in contact only with the interior wall of the flexible tube. This cleanliness feature makes it a popular type of pump for medical use, such as in dialysis.

Permanent Set

(*a*) Refers to an extension which remains in a test piece after the load has been removed, the elastic limit of the material having been exceeded.

(*b*) Refers to a permanent deflection of any structure after being subjected to a load.

Perpetual Screw. Worm.

Persian Drill. Archimedean drill.

Pet Cock (priming valve). Refers to a small plug-cock which is used for draining condensed steam from steam-engine cylinders on the starting of the engine, or for testing the water-level in a boiler.

Petrol Engine. It is an internal combustion engine which workes on the Otto 4-stroke or 2-stroke cycle using a petrol spray from a carburetter or direct petrol injection. Ignition of the combustible petrol-air mixture is done by a sparking plug which operated either by coil and battery or by magneto or by a.c., and transistor system.

Petrol Pump
    (*a*)   Refers to a small diaphragm-type pump which is operated either mechanically from the camshaft of the petrol engine, or electrically, for fuel delivery to the Carburetter.
    (*b*)   A pump at a petrol station.

Phase. Two alternating quantities are considered to be 'in phase' when their maximum values are occurring at the same instant of time.
    Phase angle. Refers to the angle between two vectors which are representing two harmonically varying quantities which have the same frequency, that is, the difference in phase measured as an angle.
    Phase difference. Refers to the difference between the phase angles of two harmonically varying quantities.
    Phase reversal. Refers to the process of getting a signal of identical waveform but of opposite phase to an original signal.
    Phase spectrum. Refers to the values of the phase angles of the components of a vibration which are arranged in the order of the frequencies.

Photoelasticity. Refers to a method which is used for determining the location and direction of stress distribution in bodies under complex systems of loading by passing polarized light through a model made of a transparent plastic material, like nitro-cellulose. The light gets polarized and transmitted only on the planes of principal stress. The stress distribution has been observed through a second piece of Polaroid material called the 'analyser'.

Physical Set-up. This term is used for describing the programming of a machine by physically fixing stops, setting switches, inserting punched cards/magnetic tapes, etc. The method has been used for automatic machines which have been not true robots.

Pick-and-place Device (loading arm). Refers to the manipulating device having a rather limited range of possible movements and a rudimentary control system. They are sometimes regarded as very simple robots.

Pick-up Well. Refers to a small petrol reservoir which provides a temporarily enriched mixture during the acceleration of an automobile and gets arranged between the metering jet and the spraying tube in some careburtters.

Pickering Governor. Refers to a governor in which the balls get connected to the centres of cambered springs of flat steel, the ends of the springs being pulled inwards as speed increases and thereby reducing the opening of a throttle valve.

Pickling. Refers to the removal of scale, grease, or salt water deposits by dipping in a suitable dilute acid circulating bath. It has been done before bounding metals with high-grade adhesive.

Pick-off. A transducer which is done used for monitoring or stabilizing a servo-mechanism. The input has been a relative displacement of its two components and the output has been generally electrical.

Pile-drawer. An appliance which is used for extracting piles from the ground.

Pile-driver

   (*a*)   Refers to a power unit which raises and lets fall a weight, called a monkey, between guides to drive in a pile.

   (*b*)   Refers to hydraulic and quiet version uses its own weight and its grip on adjacent piles either to drive in or pull out piles by hydraulic means.

Pilger Mill. A rolling mill which is used for rolling tubes, using a mandrel. Figure shows how a mill works as a discontinuous process.

Pillar Drill. See drilling machine and Fig. 3.

Pilar Pump. It is a lift pump or face pump which gets attached to a base plate carrying a pillar which forms the support for a crank, flywheel and handle for working the rod of the bucket or piston.

Pillars

   (*a*)   The lower half of a journal bearing.

   (*b*)   Bar or rods for holding apart the components of a mechanism.

Pillow Blobk (pillow). Plummer blocks.

Pilot Drill

   (*a*)   A drill which is used to produce a pilot hole .

   (*b*)   A drill which has had its pointed end concentrically reduced in diameter to precisely fit into a pilot hole thus

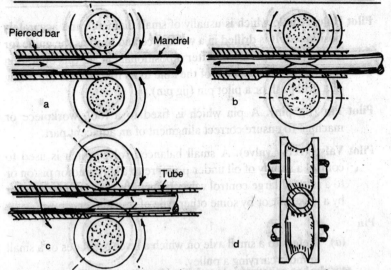

Pierced bar
Mandrel
Tube

a

b

c

d

Pilger mill, (a) insertion, (b) withdrawal, (c) rotation and advance, (d) cross-section of rolls.

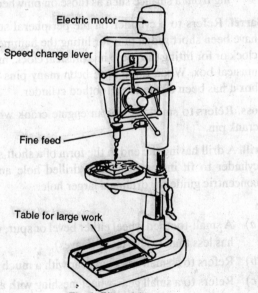

Electric motor

Speed change lever

Fine feed

Table for large work

Fig. 3. Pillar drill

enabling the accurate drilling of a large size hole from an small pilot hole.

Pilot Hole. A hole which is usually of small diameter and accurately positioned. It is drilled in a workpiece and used as the guide for a susequent operation. After a positional check, this operation may be the enlargement of the hole to its final size or the fitting of a drill bush or a pilot pin (jig pin).

Pilot Pin (jig pin). A pin which is fixed to a tool, workpiece or machine to ensure correct alingment of an adjacent part.

Pilot Valve (relay valve). A small balanced valve which is used to control a supply of oil under pressure to a survomotor piston or to a relay of large control valve. It could be operated by hand or by a governor or by some other type of transducer.

Pin

(*a*)   Refers to a small axle on which a lever oscillates or a small spindle carrying a pulley.

(*b*)   Refers to a very small-diameter cylinder which is projecting from a surface such as those on pinwheel.

Pin Barrel. Refers to a cylinder on the peripheral surface of which have been short radial pins for lifting the hammers in a chiming clock or for lifting the comb in a musical clock, musical watch or musical box. When there have been many pins as in a musical box it has been termed as a toothed cylinder.

Pin Boss. Refers to small boss of an engine crank which carries the crank pin.

Pin Drill. A drill having the end in the form of a shoft small- diameter cylinder to fit into a previously drilled hole and thus form a concentric guide for drilling a larger hole.

Pinion

(*a*)   A small-toothed wheel either bevel or spur, which normally has less than twelve teeth (leaves).

(*b*)   Refers to a small wheel in gear with a much larger one.

(*c*)   Refers to a small gear-wheel meshing with a rack.

(*d*)   Equal gears if the diameter has been equal to or smaller than the width.

Pinion Leaf. Refers to a tooth of a pinion.

Pin-jointed

    (*a*)   Said of joints in mechanisms where the only connection has been a pin about which both the joined parts can turn without restriction.

    (*b*)   Said of joints in structural frameworks in which movement have not been transmitted from one member to another.

Pinking. Knocking.

Pinwheel (pin wheel). A wheel having pins fixed a right-angles to its plane for lifting the hammer of a striking clock or repeater; also called hammer wheel.

Pinwheel Gear. It is a rotary gear. It is a rotary disc which is used for carrying an array of pins or teeth which mesh with a pinion that slides on a shaft at right-angles to the axis of disc's rotation. This constitutes a geared pair with a ratio that has been predetermined function of rotation. The rotation of the pinion shaft has been proportional to the square of the rotation of the disc when the pins are equally spaced along an Archimedean spiral.

Pinwheel Principle (odhner wheel). A principle which finds use in most barrel-type calculating mechines and illustrated in Figure 4. The essential feature has been wheel which is having nine teeth or pins that can be retracted. Each pin moved in a radial slot and is having a stud on one side which moves in a two part race cut of the setting lever. The pins get projected or retracted as the setting lever is moved, the radii of the two parts being different. The pinwheel has been geared to a number wheel in the multiplier register of the calculating machine and gets rotated it through the required number of places.

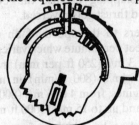

Fig. 4. Pinwheel principle.

Pipe-bending Machine. A machine which is used for bending pipes or tubes between rolls operated by a lever. The pipe has been often filled with short lengths of rod, slightly smaller than the bore, joined together and puller out after the bending operation; their insertion does not allow buckling of the pipe.

Pipe-threader. A machine which is used for cutting screw threads in metal tubing.

Piston. It is a solid or hollowed cylindrical plunger which is able to reciprocate in a cylinder either under fluid pressure in an engine or to displace or compress a fluid in pumps and compressors.

Piston which are produced by turning to be slightly oval to compensate for unequal diametral expansion associated with the material required to attached the small end of the connecting rod.

Piston Air Pump. It is a marine engine air pump having a solid piston and fitted which suction and delivery at both end.

Piston Engine. Internal-combustion engine.

Piston Pin. Gudgeon pin.

Piston Ring. It is a cast-iron ring of rectangular section which is cut through at one point to increase its springiness and to allow for fitting in a circumferential groove in the piston. The ring is springing outwards against the cylinder wall to prevent leakage.

Piston Rod. Refers to the rod which is attached to the piston of an engine or pump to transmit its motion to or from the connecting-rod or crank.

Piston Rod Gland. Refers to the gland in the stuffing box of an engine through which the piston rod passes.

Piston Slap. Refers to the slight knock which caused by a loose or worn piston slapping against the cylinder wall when the connecting-rod thrust gets reversed.

Piston Speed. Refers to the speed of an engine piston which is measured in feet per minute which varies greatly with the type of engine from 1.3 m/s (250 ft per min) in condensing, and heavy engines up to 4m/s (800 ft/min) in marine oil engines and locomotive engines, from 4 to 6 m/s (800 to 1200 ft/min) in large disel engines and up to 15 m/s (3000 ft/min) in aircraft engines.

Piston Valve. A slide valve which is formed by two short pistons attached to the valve-rod which slide over cylindrical ports in a close-fitting valve body, as in the steam chest of a steam locomotive.

Pit. See engine pit.

Pitch

(*a*)   Refers to the uniform spacing of adjacent elements of a series of points.

(*b*)   Refers to the distance between corresponding points on adjacent threads.

(*c*)   Refers to inclination or rake of the teeth of saws.

(*d*)   The angle of setting of some tools.

(*e*)   Refers to length of two links in a chain cutter.

(*f*)   Refers to the angular movement of a robot's wrist about a horizontal axis perpendicular to the axis of the robot's arm.

Pitch (helicopter). Refers to the angular setting of a helicopter blade, which is variable.

Pitch (propeller). The pitch setting refers to the blade angle measured at a standard radius, usually at 0.75 (sometimes 2/3) of the peripheral radius.

Braking pitch is a pitch which is setting to give negative thrust, including reverse pitch. Feathering pitch provides the minimum drag when the engine has been stopped.

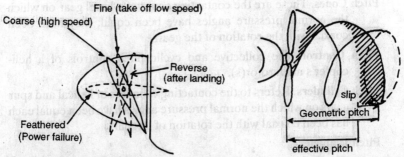

Pitch settings for variable-pitch propeller (M.E.)   Propeller pitch

Effective pitch refers to the actual distance the element moves and slip the difference between geometric and effective pitch.

Geometric pitch refers to which an element of a propeller would advance in one revolution when moving along a helix to which the line defining the blade angle of the element is tangential.

Reverse pitch. A large negative pitch setting of a propeller after an aeroplane has landed, to act as an air bracke.

Pitch Angle. Refers to the angle between the axis of a bevel gear and the pitch cone generator, being the complement of the back cone angle.

Pitch Chain (sprocket chain). It is chain flat links between whose sides the projections of a sprocket wheel engage.

Pitch Circle

(*a*)   Refers to the circumference of the pitch line. For two wheels in mesh, the pich circles roll in contact.

(*b*)   Refers to circle of intersection of the pitch cone of a bevel gear and the outer end faces of the teeth.

Pitch Cones. These are the contacting cones of a bevel gear on which the normal pressure angles have been equal; they have been coaxial with the rotation of the gears.

Pitch Control. The collective and cyclic pitch controls of a helicopter's main rotor(s).

Pitch Cylinders. Refers to the contacting cylinders of helical and spur gears on which the normal pressure angles have been equal each has been coaxial with the rotation of its gears.

Pitch Diameter

(*a*)   Refers to the diameter of the pitch circle or cylinder of a gear.

(*b*)   Refers to the diameter pitch cylinder of a parallel screw thread or the nominal diameter of the pitch cone of a tapor screw thread.

Pitch Plane. A plane in axially toothed worm wheels which is parallel to both the axes of the worm and worm wheel and tangential to the pitch cylinder of the worm wheel.

Pitch Point

(*a*) The point of contact of a pair of pitch circle of gear.

(*b*) Refers to the point where the pitch line intersects the flank of a screw thread.

Pitch Surfaces. Refers to surfaces that roll together with no sliding like the pitch cones of bevel gears and the pitch cylinders of spur and helical gears.

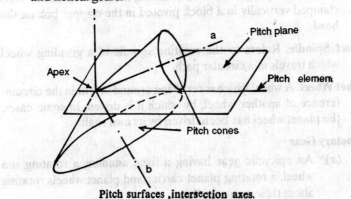

Pitch surfaces, intersection axes.

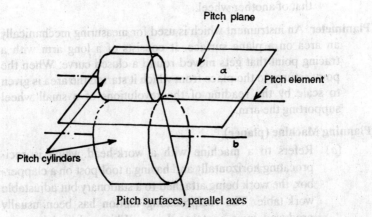

Pitch surfaces, parallel axes

Pit-head Gear. All the machinery and its framework which are erected over a pit's mouth for raising and lowering the case.

Pit Wheel. A mortise wheel which is revolving in a pit on a horizontal axis. It is usually the first motion wheel in a water or wind-mill.

Pivot

    (*a*) Refers to a short shaft or pin on which something turns oscillates.

    (*b*) Refers to the end of an arbor which turns in a hole, jewel or screw in a clock or watch. The end can be parallel, shouldered or conical.

Planner Tools. Cutting tools for a planning machine which has been clamped vertically in a block pivoted in the clapper box on the head.

Planet Spindle. Refers to the rotating spindle of a grinding wheel which travels in a circular path.

Planet Wheel. A wheel which is revolving around or within the circumference of another wheel, by which it is driven. In some cases, the planet wheel has been driven by its own shaft.

Planetary Gear

    (*a*) An epicyclic gear having a fixed annulus, a rotating sun wheel, a rotating planet carrier and planet wheels rotating about their own spindles.

    (*b*) Any gearwheel whose axis describes a circular path round that of another wheel.

Planimeter. An instrument which is used for measuring mechanically an area on a plane surface. It consists of a long arm with a tracing point that gets moved round a closed curve. When the point returns to the place from which it started the area is given to scale by the reading of the revolutions of a small wheel supporting the arm.

Planning Machine (planer)

    (*a*) Refers to a machine with a work-head which is reciprocating horizontally and having a tool-post on a clapper-box, the work being attached to a stationary but adjustable work-table. The reciprocating motion has been usually produced by a scotch yoke or Whitworth quick-return motion.

    (*b*) Refers to a machine which consists of a gear-driven reciprocating work-table sliding on a heavy bed with a stationary tool above the table on a saddle. It can be traversed across a horizonal rail carried by uprights for producing large flat surfaces.

Planoid Gears. Hypoid-type gearing with the pinion offset limited to one-sixth to one-third of the gear diameter. They are used in 1.5 to 1 up to 10 to 1 velocity ratio range.

Planometer. See surface plate.

Plant. All the machinery requisite which is used for carrying on business in a factory. (Also sometimes the building and site).

Plastic Deformation. Deformation while a material has been in a plastic state.

Plastic State. See stress-strain relationship.

Plasticity. This property is used to refer to the susceptability of a material at a certain temperature and loading condition to exhibit a permanent deformation after is gets induced by a stress taking it past the yield point.

Plate. A term which is usually applied to flat metallic materials greater than 5mm (3/16 in) thick.

Plate-beading Machine. A machine which consists of three rolls with bearings in housings, the top roll being adjustable to vary the curve when binding boiler and other plates between it and the other two below.

Plate Clutch. See clutch.

Plate-edge Planning Machine. Refers to a metal-planning machine which is used for truing the edges of plates, having a fixed table and a travelling tool.

Plate-flattening Machine. A straightening machine which is using seven rolls, four above and three below.

Plate-gauge
    (*a*)   A thin, flat metal gauge which is used for measuring spaces.
    (*b*)   An external gauge which is formed by cutting slots of the required gauge width in a steel plate, the surfaces of which are hardened.

Plate Link Chain. See chain.

Plate Mill (plate rolls). Refers to a rolling mill having rolls which are plain cylinders, Grains rolls do the roughing and chilled rolls the finishing process for the plates.

Plate Rolls. See plate mill.

Plate-shearings Machine. A shearing machine which is furnished with specially long shears. It is used for cutting off the ragged edges of plates after leaving the rolls.

Platen. Refers to the work-table of a machine-tool which is usually slotted for clamping the-headed bolts.

Platen Machine. A printing machine which is using a flat surface.

Plates

    (*a*)   Thick sheets of metal.

    (*b*)   The brass plates which form the framework of a clock or watch and in which holes are drilled for the pivots of the train of wheels.

Platform Scale. A small weighbridge.

Play. Refers to a limited freedom of movement in a bearing or a working part of a mechanism.

Plug

    (*a*)   The inner movable portion of a cock (*b*).

    (*b*)   A small arbor or chuck which is used in a larger lathe chuck.

Plug Cock. See plug (*a*).

Plug Cushion Process. This method of explosive forming is shows in Figure.

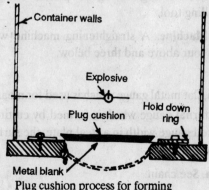

Plug cushion process for forming

Plug Gauge. Refers to a gauge which is used to check the dimension of a hole.

Plug Rod. Plug tree.

Plug Tree (plug rod). Refers to a long rod which is suspended from the beam of a single-acting pumping engine and provided with tappets for moving the handles of the equilibrium and steam exhaust valves.

Plummer Block (pillow block). A box-form casing which holds the brasses or other bearing metal for a journal bearing on line shafting and split horizontally to take up wear. The base of bearing has been sometimes termed the 'pillow' and hence the term 'pillow block'. Lubrication of the bearing has been usually by an oiling ring.

Plunger

(a) Refers to the solid piston or ram of a force pump. A plunger is different from a piston in being longer than its stroke.

(b) Refers to the solid piston which is used in moulding and extrusion.

Plunger Bucket. A force pump which is having no valves.

Plunger Pump. Force pump.

Pneumatic Brake. Refers to a continuous braking system in which air pressure has been applied to brake cylinders throughout a train.

Pneumatic Conveyor. See pneumatic tube conveyor.

Pneumatic Drill. It is a rock drill which uses compressed air to reciprocate a loose piston which hammers the shank of the bit or an intermediate piece, or the bit is clamped to the piston rod. Provision is made to rotate the bit by a small amount between each stroke in most models.

Pneumatic Hoist. A light hoist which is used car lifting loads directly by the movement of a piston in a long cylinder suspended over the work. It is operated by compressed air.

Pneumatic Pick. Refers to a straight pick which gets hammered rapidly by a reciprocating piston driven by compressed air.

Pneumatic Riveter. It is a high-speed riveting machine in which a rapidly reciprocating piston has been driven by compressed air delivering up to 2000 blows per minute.

Pneumatic Tools. Tools which are driven by compressed air.

Pneumatic Tube Conveyor. Refers to the conveyance of small objects in suitable containers along tubes either by a vacuum or by air pressure.

Point Bar. A horizontal bar which is supporting the points (b) at the back and front of a lace machine and moving with the swing of the carriage.

Point Chuck. Refers to the point centre which gets attached to the headstock of the lathe for turning work which is pivoted between centres.

Point Ground. The set of a tool, such as a drill, which normal helix angle has been ground locally at the tip to change the form of the cutting edge, as for cutting stainless steel.

Poise (P). It is a unit of dynamic viscosity and is usually quoted in centipoise (cP); 1 cP = 10^{-3} Ns/m^2.

Poission's Ratio. Refers to the ratio of lateral contraction per unit breath to longitudinal extension per unit length when a piece of material gets stretched or compressed. The tensile and compressive values could slightly.

Polar Configuration. The configuration of a robot having one transtational and two rotational degrees of freedom. The working space forms part of a sphere.

Polar Moment of Inertia. May be defined as the moment of intertia about an axis through the centre of gravity which is perpendicular to the plane of the figure. It has been equal to the sum of the moments of intertia about two perpendicular axes through the centre of gravity in the plane of the figure.

Polishing. The term used for making smooth and glossy, usually by friction, such as by a polishing wheel or mop to remove irregularities resulting from machine operations.

Polishing Lathe. A lathe fitted with arbor wheels, mops etc., which is used for polishing.

Poncelot Wheel. An undershot wheel having curved instead of the more usual flat vanes, and consequently more efficient.

Poppet (poppet head). Refers to the moveable headstock of a lathe, the design of which allows the work to get revolved between centres.

Poppet Cylinder. Refers to the cylindrical poppet mandrel of a lathe.

Poppet Valve (mushroom valve). A mushroom-shaped valve which is commonly used for inlet and exhaust valves of an internal-combustion engine. It consists of a circular head with a conical face which seats in a conical port in the cylinder and has a guided steam by which it gets lifted, using a rocker arm and/or tappet.

Port. Refers to a cylinder opening in an engine, pump etc., by which a fluid enters or leaves, usually under the control of a valve.

Porter Governor. Refers to a pendulum governor in which the ends of two arms have been pivoted to the spindle and sleeve respectively and carry heavy balls at their pivoted joints.

Position Gauge. A gauge which is used for checking geometrical relationships within assigned tolerances.

Positive Movement. Refers to a movement in any part of a loom due to mechanical means.

Potential Energy. See energy.

Poundal. Refers to a force which produces in a mass of one pound an acceleration of one foot per second. It is the unit of force in the foot-pound-second system of units, 32.2 poundals equals one pound weight.

Powder Spraying. See metal spraying.

Power. Refers to the rate of doing work.

Power Drag-line. Refers to an excavator comprising a large scraper pan or bucket which gets dragged through the material towards the machine and below its boom or jib.

Power Hammer. Refers to any type of hammer which is operated continuously or intermittently, by some souce of power.

Power House. Refers to any building in which power gets generated for distribution or for conversion and latter distribution.

Power Plant

    (*a*) A number of power, units which are assembled in one place.

    (*b*) The complete propulsive unit for vehicle, especially an aircraft.

Power Rating or Horsepower. Refers to the output power of a motor or aero engine which is measured in watts or horsepower under specified conditions including torque and rad/s or rew/min. For aero piston engines we use the manifold pressure or boost pressure and the torque; for subjects and turboprops the jet-pipe temperature and torque; and for rocket motors the pressure in the combustion chamber.

Power Shovel (nevvy). It is an excavator which consists of a jib with a radial arm along which a large buket or scoop travels. The bucket is making a radial cut and digs above the level of the excavator.

Power Unit. The term used for an engine, or assembly of engines, complete with shafts, gears, etc., and in the case of aircraft including the propellers. It is often termed as power plant.

Preadmission. The term used for the admission of steam to an engine-cylinder just prior to the termination of a stroke in the opposite direction.

Precession. The effect on a constantly rotating gyroscope of applying a constant torque to alter the direction of its rotating axis has been that the axis slowly generates a cone. This is termed as a precessional motion, and in this example a nutation.

Preloaded Ball Screw. Refers to a combination of two ball joints which are joined together by a shimming material whose thickness gets adjusted so that there has been no backlash. Power is transmitted between the screw and the nuts through the balls which circulate through the tube to from a continuous path.

Pre-ignition. Refers to the premature firing of the explosive mixture in a cylinder of an internal-combustion engine.

Preoptive Lathe. Refers to a lathe with a special headstock which makes instaneous change of spindle speed, while cutting is in progress, by using multi-disc friction clutches.

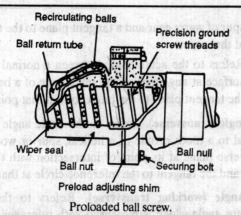

Recirculating balls
Ball return tube
Precision ground screw threads
Wiper seal
Ball null
Ball nut
Securing bolt
Preload adjusting shim

Proloaded ball screw.

Pre-release. Refers to the opening of a steam cylinder to exhaust just before the termination of the piston stroke.

Pre-selector Gearbox. Refers to the selection of a gear-ratio in the gearbox, before requirement, by the movement of a small lever. The gear has been afterwards engaged by pressure on a pedal.

Pre-tension. Refers to the amount of tensile load applied to a bolt or tie-rod when it gets installed, but not subjected to its working environment.

Press. See hydraulic press.

> *Double-acting press.* A press having two slides to permit two separate operations such as blanking and drawing to be carried out at a stroke.

Press Forging. Forging in which pressure has been applied by squeezing rather than by a hammer blow.

Press Tool. A tool which is used for forming, assembling or cutting using a punch or die.

Pressing (stamping). Refers to the production of forged work by pressing dies together under a hammer or press.

Pressure Angle (axial). Refers to the accute angle which is measured in an axial plane between the axis of a helical or worm gear and a normal to the tooth profile at a point on the reference cylinder.

Pressure Angle (normal)

> (*a*) Refers to the acute angle between a radial line which is passing through any point on the tooth surface of a helical

spur of warm gear and a tangent plane to the tooth surface at that point.

(b)   Refers to the acute angle between a normal to the tooth surface at any point on the pitch cone of a bevel gear and the tangent plane to the pitch cone at that point.

Pressure Angle (transverse). Refers to the acute angle between the normal to a tooth profile of a helical, spur or worm gear in a transverse plane at its point of intersection with the reference circle and the tangent to the reference circle at that point.

Pressure Angle (working transverse). Refers to the transverse pressure angle which is measured with reference to the pitch circle instead of the reference circle.

Pressure Differential. Refers to the difference in pressure between two chambers which is used as a driving force.

Pressure Feedback Unit. Refers to a device which is used for feeding back a displacement or other quantity proportional to the operating pressure in an hydraulic system.

Pressure Forging. Drop forging.

Pressure Gauge. Bourdon gauge.

Pressure Jet

(a)   A small jet-propulsion unit which is fitted to the tips of the rotor blades of some rotocraft.

(b)   A small jet nozzle which is operated from a gas source a board an artificial sattelite or a rocket and used to control its orientation.

Pressure Ratio

(a)   Refers to the ratio of the pressure at the beginning of a process to that on completion, at in a piston- engine for the mixture before and after compression by the piston in the cylinder.

(b)   Refers to the absolute air pressure prior combustion in a jet-engine divided by the ambient pressure.

(c)   Refers to the ratio of air (gas) pressures across a compressor, a turbine or a propelling nozzle of a jet-engine.

Pressure Roller. A roller which is used in centreless ginding of short, heavy workpieces to assist their rotation.

Prime Mover. An engine or mechanism which is able to convert a natural source of energy into mechanical power.

Priming

(*a*) Refers to the operation of filling a pump intake with fluid to expel the air.

(*b*) Refers to the operation of injecting petrol into an engine cylinder to assist starting.

Priming Pump. A fuel pump which is used for supplying a piston engine with fuel during starting.

Principal Axes. Every cross section of a beam is having two principal axes which are passing through the centre of gravity and they have been always at right angles to each other. The moment of inertia in the plane of the section refer to a maximum about one axis and a minimum about the other. For symmetrical sections, the axes of symmetry have been always the principal axes.

Principal Moment of Inertia. Refers to the moment of inertia about any of the principal axes.

Principal Planes of Stress. If a piece of material has been subjected to loadings of any kind, producing the full combination of the longitudinal and lateral tensile or compressive stresses combined with complimentary shear stress, there still exist, within the material, orthogonal planes, the P-planes, where the shear stress has been zero. In the P-planes a pure direct stress exists. The maximum shear stress exists at 45° to these principal planes.

When one of the principal stress has been zero, the condition has been one of plane stress. When two principal stresses are zero then a uniaxial stress exists.

Principal of Least Constraint. This principal stress that motions shall be such that they produce minimum constraints. The constraints, of any number of interconnected masses have been the product of each mass and the square of its deviation form its position of it were free.

Printing Machine. A machine which is used for transferring ink, etc., to paper:

(a) relief type, using rotating cylinders having raised type or a flat surface;

(b) planographic type, using prepared metal surfaces that accept or reject the ink;

(c) intaglio and photogravure types, using surfaces with ink-carrying hollowed-out portions;

(d) direct photographic process.

Process Annealing. Heating a material, like steel or aluminium in sheet or wire form, between cold-working operations, like stretch forming or rolling, at just below its critical temperature followed by slow cooling to improve its structural properties.

Process Control. The term used for the automatic control of sections of an industrial plant by electronic means; including rates and accelerations of flow plus changes in temperature, pressure etc.

Process Engineering. See process control.

Product Register. Refers to the register on a calculating machine, recording the results if multiplication and mounted on the movable engine.

Proell Governor. Refers to a type of Porter governor in which the balls get attached to upward prolongations of the links to the governor sleeve.

Profile. Refers to the shade of a normal section through a surface.

   Modified profile. Refers to a profile, excluding all texture waviness which exceeds a certain maximum.

Profile Grinding (form grinding). The term used for the grinding of cylindrical work without traversing the wheel, periphery of which is having the required profile.

Profile Milling Machine. Refers to a milling machine in which the rotating spindle gets constrained to run over a copy of the article to be milled.

Profile-turning Slide (bevel turning slide). A tool slide which is mounted on the cross-slide of a lathe. The movement of the tool in the axial direction of the lathe has been controlled by a cam fixed to the tailstock.

Profiling. The term used for the grinding and sharpening of a cutter so that it will have the correct sharp for forming the material which it has been going to cut.

Profilometer. An instrument which is used for ascertaining the quality of surface finish, using a stylus and observing the surface irregularities on an oscillograph.

Programming Arm. Refers to a lightweight structure which is used in teaching instead of using a robot's arm.

Projection (American, English, first-angle, third-angle). Standard drawing methods are used for projecting a three- dimensional object on a flat surface. The third-angle is now widely adopted and the views shown are those that could be seen by looking at the near side of the adjacent view, following the theory and conventions of orthographic projection. Other projections like isometric or oblique projections have been rarely used for manufacturing drawing except as illustrations for grain-flow or as an indication of the overall shape or in the defining where the part fits into an assembly.

Projection Gauge. Refers to an optical gauge which magnifies the profiles of screw threads about 100 to 200 times to case inspection and comparison.

Prony Brake. It is an absorption dynamometer where the torque of an engine has been absorbed by a pair of friction blocks bolted together across a brake drum and gets balanced by weights at the end of an arm attached to the blocks; alternatively, the arm has been secured to a band encircling the flywheel or pulley of the engine.

Proof Load (test load)

    (*a*)   A load which is greater than the working load to which a structure or a mechanism has been tested to ascertain whether it can withstand like a load without permanent distortion or damage.

    (*b*)   The product of the limit load and the 'proof factor of safety'.

Proof Stress. In metals which do not exhibit a sudden yield point, it refers to the stress which is required to produce a certain amount of extension.

Propeller. A power-driven bladed screw which has been designed to produce thrust by its rotation in air, water or other gas or fluid. In air often termed as an 'airscrew' and in a 'marine screw propeller'. The screw produces the thrust by providing moment-um to the column of air (water) which it drives backwards.

When the blades have been fixed for an aircraft propeller it has been termed as a 'fixed' pitch propeller', but when the setting can be changed it has been termed as adjustable pitch or a variable pitch propeller.

Cycloidal propeller. A marine propeller has four identical hydrofoil shapes (Figure 12) which are rotating about a vertical axis. With N and O coincident there has been no thrust. When N gets displaced to alter the pitch cyclically the direction of thrust is changed as shown.

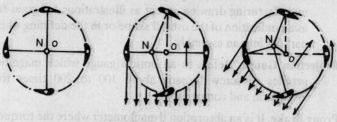

Cycloidal propeller.

Propeller Efficiency. Refers to the ratio of the thrust of a propeller to the torque supplied by the engine shaft. It is generally 75.80% and up to 90% in special cases for aeroplanes.

Propeller Fan (fan). Refers to an impeller or rotor which is usually fitted with blades of aerofoil form working in a cylindrical casing to provide a current of air.

Propeller Shaft. Refers to the driving shaft which conveys the engine power from the gear-box to the rear or front axle of a motor-vehicle. It is usually connected through universal joints to allow for vertical displacement of the axle on the springs. Sometimes termed as 'cardan shaft'.

Propeller-type Water Turbine. A water turbine which is having a runner similar to a four-bladed marine propeller.

Propelling Nozzle. The nozzle which gets attached to the rear end of a jet-pipe, or to an exhaust cone or to the rear end of a rocket-engine. It is usually a fixed throat of convergent-divergent type, the area of which has been sometimes varied by the movement of a central bullet or by a slotted shroud called an 'ejector exhaust nozzle'.

Proud. Refers to a portion of work-piece which is projecting beyond its surroundings.

Proving Machine. Testing machine.

Puddling Rolls. Refers to the first set of rolls through which a shingled bloom is passed.

Pug Mill. A mill which is used for mixing concrete ingredients.

Pulley. A wheel on a shaft with either a cambered rim for carrying an endless belt or grooved for carrying a rope, vee-belt or chain.

Pulley Block. Pulleys which are placed side by side in a wooden or metal frame.

Pulley Lathe. A machine which is used for boring and facing pulleys having special arrangements for setting and turning.

Pull-over Mill. It is a rolling mill having a single pair of rolls so that, after passing through the rolls, the metal is to be pulled back over the top roll for the next feed.

Pulsator. It is an apparatus which is used for causing alternate suction and pressure release fifty to sixty times a minute, as used with milking machines.

Pulsejet Engine. It is an air-swallowing engine which is composed of a combustion chamber to which air gets admitted through valves that are opened or shut by the pressure in the chamber, and of a nozzle which generates thrust by a jet of hot gases. A sparking-plug is needed to ignite the fuel to start the engine but thereafter the operation is automatic. The successful functioning of this resonating type of engine is dependent on the proper matching of the natural frequencies of the pressure and expansion waves within the combustion system and the mechanical properties of the valves.

Pulsometer (pump). Refers to a steam-condensing vacuum pump which is so-called from the pulsatory action of the steam, having an automatic ball-valve as the only moving part admitting steam alternately to a pair of chambers. It is sometimes used for dealing with liquids which are having solid matters in suspension.

Pump. A mechanism which is used for converting mechanical energy into energy in a fluid.

Phase pump. A pump having two opposing pistons which are working in each cylinder with a variable phase of operation.

Redial pump. A pump having the cylinders radially disposed.

Replenishment pump. A pump which is delivering fluid at a suitable pressure for replenishing a system. When the pressure in the low-pressure line is raised above ambient pressure it is called 'forced replenishment'.

Swash-plate pump. A pump having axial cylinders and pistons (or their connecting-rods) on an inclined member, the pistons reciprocating when there has been relative motion between the inclined member and the cylinder system.

Vane pump. A pump having its rotor axis which is mounted eccentrically in a cylindrical pressure chamber and with a number of vanes in radial slots maintained in contact with the chamber surfaces by springs or fluid pressure. The movement of the rotor makes delivery of the fluid.

Variable-stroke pump. A radial pump or swash-plate pump having a variable crank throw or swash angle. If varied automatically as when maintaining a constant pressure it is termed as an 'auto variable-stroke pump'.

Pump Barrel. Refers to the closed cylinder in which the bucket, plunger or piston of a pump moves.

Pump Bucket. It is the piston of plunger of a suction pump.

Pump Duty. Refers to the overall efficiency of a pump which is usually measured as the ratio of mechanical work output to heat input.

Punch

(*a*) A non-rotating steel tool which is used for shearing out a piece of material of a certain shape under pressure from plates, etc. It is supported underneath by a die with a slightly larger profile of the same shape.

(*b*) A tool which is used in extrusion.

(*c*) A sharp pointed tool which is used to mark the centre points of holes before they get drilled in a workpiece.

Punching Bear. A portable punching machine.

Punching Machine. A machine which is used for punching holes in plates using a punch driven by a crank and reciprocating block or by a hydraulic ram.

Purchase

(*a*) Mechanical advantage or leverage.

(*b*) A mechanical appliance which is used for gaining a mechanical advantage.

Pushrod and Rocker. A rod which is operating the valve rocker of an overhead- valve engine when the camshaft gets located in the crankcase. The vertical rod under H in has been a pushrod'.

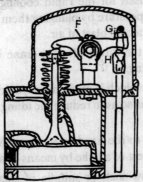

Pushrod and rocker. (F) Valve rocker. (G) Tappet clearance adjusting screw. (H) Spherical cup.

Pusher (aeroplane). Refers to a piston-engine aeroplane in which the engine is producing compression in the propeller shaft.

QTAT. Quick Turn Around Time.

Quadrant Plate (wheel plate). The plate which is carrying the stud wheels in the gear-box of a screw cutting lathe.

Quadruple-expansion Engine. Refers to steam-engine in which the steam gets expanded successively in four cylinders of increasing size working on the same crankshaft.

Quartering. Refers to the adjustment of cranks or crank pinholes at right-angles to each other.

Quartering Machine. A double-headed machine which is used for accurate boring at right angles of crankpin holes in locomotive driving wheels after fixing on their axles.

Quenching. The term used for the rapid cooling of steels from an above-critical temperature by plunging them into water, oil salt, a molten metal or a cold blast of air.

Quick Gear. Refers to the direct lift of a crane instead of through intermediate gearing.

Quick Return. A return which is made more rapidly than the cutting stroke of a machine, so as to reduce the idling (non-cutting) time of the tool.

Quick Traverse. Refers to the hand or mechanical traverse which is given to the slide rest of a lathe by means of a rack and pinion.

Quill. Refers to a hollow shaft or spindle of small diameter which damps out the effect of starting and compensates for misalignment with the pinion shafting due to applied load; it can be a weak link when it is necessary to prevent overspeeding of the driven part, as with an aeroplane propeller.

Quill Drive. Transmission through a quill. It is a drive through a hollow shaft concentric with a solid shaft, but on separate bearings. The shafts can get connected by a clutch or by a flange at one end to provide a flexible drive. In a geared quill drive the quill is carrying a gear-wheel which gets geared to a pinion on a shaft.

R

RAM. Random Access Memory within a digital computer or robot.

RF. Reserve factor.

Ref. Mp. Referred Mean-pressure.

Rev/min, (rpm). Revolutions per minute.

Rev/s (rps). Revolutions per second.

Race

    (*a*)   Refers to the inner or outer steel rings of a ball-bearing or a roller-bearing.

    (*b*)   Refers to a circular ring which is supporting a revolving superstructure that moves on rollers, as in some cranes.

Race Board (shuttle race). Refers to that part of the sley in a loom along which the shuttle travels.

Rack

    (*a*)   Refers to a straight length of toothed gearing.

    (*b*)   Refers to a bar rail having teeth for engaging with a moving mechanism by means of a geared wheel; a cogged rail. A helical rack is shown in

    (*c*)   Refers to a straight bar or a flat plate having a series of equi-distant teeth on one face (see also rack and pinion) When the pitch radius of a spur or helical gear becomes infinite, it is a rack.

Rack and Pinion. Refers to an arrangement of a straight-toothed rack and a pinion which is used to convert rotary into linear and reciprocal motion, generally a pinion wheel with fixed centre actuating a movable rack.

Rack and Pinion Steering-gear. Refers to a steering-gear for motor-vehicles in which a pinion on the steering-column gets engaged with a rack attached to a divided trackrod.

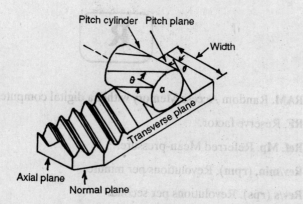

Helical rack nomenclature.

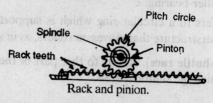

Rack and pinion.

Rack Double-sided. A rack which is having teeth machined on opposite sides of a straight bar.

Rack Feed. Feed by a rack and pinion.

Rack, Helical. Refers to rack with the pitch plane for its pitch surface and its axial and transverse directions those of the mating gear.

Rack Railway. Refers to a railway on a steep gradient in mountainous districts which is using locomotives with toothed gear to engage a rack or cogged rail, and thus obtain additional adhesion. The rack is generally laid in the centre between the rails.

Rack Spur. A spur gear having an infinitely large pitch diameter.

Racking. The term used for moving a machine mechanism, or part thereof, forwards and backwards.

Racking Gear. Refers to the gear which is operating the block carriage of a crane.

Radial Engine. Refers to a piston-engine having the cylinders arranged radially at equal angular intervals around the crankshaft. They have been either in one plane (single-row) or in two planes, one behind the other (double-row).

Radial Paddle Wheel. Refers to a paddle wheel with fixed float boards which is fastened directly to the arms with their faces radiating from the centre of the wheel.

Radial Valvegear. Refers to the valvegear of a steam-engine in which the slide valve has been given independent component motions proportional to the sine and cosine of the crank angle.

Radian (rad). May be defined as the angle which is subtended at the centre of a circle by an arc whose length is equal to the radius of the circle.

Radiator. Refers to a device for cooling a relatively hot body. Usually this is attained not by radiation but, by conduction of heat into a cooler fluid passing between radiator fins. Frequently the exchange of heat takes place between two fluids liquid/luquid; liquid metal/liquid; liquid/gas, or gas/gas.

Radius of Gyration. May be defined as the square root of the ratio of the moment of inertia of a body about a given axis divided by its mass.

Radius Rod (brindle rod)
  (a) Refers to a rod attached to the die or block of a Walsvchaert's valve-gear which is used for transmitting its motion to the end of the combination lever pivoted to the valve rod.
  (b) A rod which is passing from the die block of a slot link to the slide or cut-off valve of a steam engine.
  (c) A rod in a parallel motion arrangement, fixed on a pivot at one end and jointed to the back link of the other.

Rake (rake angle)
  (a) May be defined as an angle of inclination or an angular relief 'Front rake, 'side rake'. 'top rake' are positional names assigned to the faces of cutting tools whose angles are set to obtain the most efficient cutting angle.
  (b) The pitch of saw teeth.

Back rack. For a turning tool, it refers to the angle the top, working surface makes with a plane parallel to the stem of the tool and therefore the machine's tool cross-traverse direction.

Positive rake. A tool cutting inclination which makes the working face have an obtuse angle between itself and the work-piece.

Negative rake. A tool upper face having inclination of less than a right angle.

Ram

(a) A hydraulic ram.

(b) The arm of a shaping or slotting machine which is carrying the tool backwards and forwards.

(c) A term which is used to describe the effect of the speed of an aircraft on the induced pressure in the engine air intake.

Ram-air Turbine (wind-driven turbine). Refers to a small turbine in an exposed position on an aircraft which is driven by the ram air to provide a source of auxiliary power for electric generators, fuel and hydraulic pumps etc.

Ramjet Engine. Refers to an air-swallowing engine which is composed of a diffuser, a combustion chamber and a nozzle, which generates thrust by a jet of hot gases. It has no moving parts except the fuel pump.

Ramps (guide plates)

(a) Inclined surfaces.

(b) Appliances which clip on to the rails and are provided with flat helical extension so that the wheels of rolling stock can slide up and on to the rail.

Rankine Cycle. Refers to a cycle which is used as a standard of efficiency for composite steam plants. It is comprising the introduction of water at boiler pressure, evaporation, adiabatic expansion to condenser pressure, and condensation of the steam to the initial point.

Rankine Efficiency. Refers to the efficiency of an engine when compared with an ideal engine working on the Rankine cycle under given conditions of steam temperature and pressure.

Ratchet (ratchet wheel)

(*a*) A wheel having inclined teeth for engaging having a pawl which allows only forward motion an arrests backward running.

(*b*) A wheel having specially shaped pointed teeth which engages having a click, as used on a barrel arbor in watches and clocks to prevent it turning back when the spring is being wound.

Ratchet Jack. A screw jack which is worked by means of a ratchet and pawl.

Ratchet and double-ended pawl.

Ratchet Pawl. A pawl which engages with a ratchet wheel or a spur wheel.

Ratchet Teeth. Fine-pointed teeth on an escape wheel.

Ratchet Wheel. Racthet.

Rate Gyro. See gyro.

Rated Altitude. Refers to the altitude at which a piston-engine provides its maximum power. With more than one stage of supercharging there will be correspondingly, more than one rated altitude.

Ratio of Compression. Compression ratio.

Ratio of Expansion. Expansion ratio.

Reaction Turbine. The term used for a turbine in which the working gas expands through alternate rows of stator and rotor blades.

The rotor blades absorb the kinetic energy after each expansion through a set of stator blades as shown simplicitally.

Figure shows a turbine rotor unit as found in a turbojet engine.

Reaction Wheel. An enclosed wheel actuated by water pressure as in a turbine.

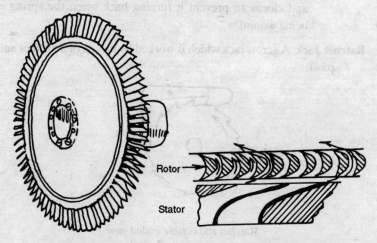

Francis water turbine.   Turbine rotor unit.

Reamer (Rimer). A tool with longitudinal or spiral flutes or separate teeth on a cylindrical or conical shank for finishing drilled holes.

Reaming. Enlarging a hole in metal with reamer.

Reaper (reaping machine). An agricultural machine for cutting grain which is having a protruding arm with fixed knives (fingers) and a moving seythe working between the slots of the fingers and actuated by gearing from the wheel of the machine. Usually there has been a paddle wheel to cause the cut stocks to lie all in the same direction.

Reboring. Reboring a worn cylinder and fitting a slightly larger diameter piston.

Rebound Hardness. Dynamic hardness.

Receiver Gauge. A gauge which is used for checking simultaneously all relevant features of a component.

Recess (Faucet). Refers to a depression or hollow on a surface into which another part can be fitted.

Rechucking. Resetting a piece of work in a lathe chuck to complete turning, etc.

Reciprocating Engine. Refers to an engine with a piston which is oscillating in a cylinder under the periodic pressure of the working fluid.

Reducing Valve. A valve which is used for reducing the pressure of a fluid in a supply line.

Reduction Gearing. Gearing or a system of pulleys to apply a source of power at a lower rotational speed (rev/min).

Redundant Structure. A structure which is having more members than are necessary to make it rigid.

Reduplication. Refers to the gain in power from a combination of pulleys in pulley blocks.

Reeling. In general, winding on to a reel.

Reference Gauge. A gauge which is used for reference to check other gauges or of the final product.

Reflux Valve. Check valve.

Regulator

    (*a*)   Chronometer.

    (*b*)   The index.

    (*c*)   A gas regulator.

Regulator Valve. Refers to a valve for controlling the steam supply to the cylinders of a locomotive, operated from the cab and located in the dome or steam space.

Reheat. Refers to the burning of additional fuel after the turbine of a turbojet engine to provide additional thrust. Also called 'after burning'.

Reheating (re-superheating). Refers to the passing of partially expanded steam in a steam-turbine back to a superheater before its further expansion.

Relative Efficiency. Refers to the ratio of the actual indicated thermal efficiency of an ideal cycle of an internal combustion engine. An air standard at the same compression ratio is one such ideal cycle.

Relay. A device which is used for supplying additional energy, such as more pressure.

    Hydraulic relay. Hydraulic means by which a mechanical displacement gets raised from a small power level to another displacement at a higher power level.

    Jet relay. A hydraulic relay in which the momentum of a fluid jet provides the necessary force.

    Pneumatic relay. Pneumatic means by which a mechanical displacement is raised from a small power level to another displacement at a higher power level.

Release

    (*a*)   Refers to the operating of the steam port of a steam-engine to allow the escape of the exhaust steam.

    (*b*)   A trigger arrangement which is used for releasing the shutter of a camera.

Relief Valve

    (*a*)   A safety valve.

    (*b*)   A cylinder escape valve.

Relieving. Backing-off.

Relighting. Restarting a gas-turbine engine or reheat in flight.

Remote Centre Compliance. Type of compliance mechanism for robot end effectors especially during insertion operations. It permits the object being inserted to rotate about its tip, as when inserting a screw.

Repeatability. Refers to the minimum tolerance achievable by a robot returning to a point in space visted before under indentical conditions.

Reserve Factor (factor of safety). Refers to the ratio of the stress (or load) that would cause failure to the stress (or load) that is experienced in service.

Resilience. Refers to the stored energy of a strained or elastic material, such as in a compressed spring or in rubber dampers which have inherent damping properties.

Resolution. Refers to the smallest achievable incremental step of a robot.

Retaining Valve. An additional valve in a pump which is used to prevent water running back when the lift is from a great depth.

Retarder (skate, wagon retarder). An arrangement of braking surfaces, parallel and alongside the running rails in a shunting yard, which have been operable from a single-box.

Return Crank. Refers to the short crank which is fixed to the outer end of the main crank pin, that replaces an eccentric in Walschaert's valve gear on locomotives with outside cylinders.

Return Valve. An overflow valve, which allows the return of fluid.

Reverse-flow Compressor. Refers to an axial-flow compressor with the axial flow in a forward direction as in some turboprop engines, the flow being reversed before entering the combustion chamber.

Reverse Gear. Refers to the gear by which an engine provides power for movement in a direction opposite to the normal direction.

Reverse Jaw Chuch. A dog chuck whose jaws have been reversible, end for end.

Reverse Jaws. Refers to the jaws of a lathe chuck which are placed within the work when turning exterior surfaces.

Reverse Keys. Refers to a pair of steel plates, one with a projecting slip on one edge and the other with a recess of the same length, which have been used as a wedge to drive two machine parts away from each other.

Reversing Cam. A cam operating the valves of a gas-engine, which has been arranged to shift along a shaft, or have its motion reversed, in order to make the engine run in the reverse direction.

Reversing Countershaft. Refers to a countershaft whose direction of rotation can get reversed.

Reversing Engine. An engine whose direction of motion can get reversed.

Reversing Gear. Refers to the gear which reverses the direction of motion of an engine machine or mechanism.

Reversing Link. Refers to the slotted link of an engine which alters a valve for forward or backward motion.

Reversing Plate. A plate keyed on a crankshaft and furnished with a slot, or holes, by which an eccentric is moved up or down for forward or backward motion respectively. It is found on a single-cylinder engine with not slot link or for operating a valve opening.

Reynolds Number, Re. Refers to the product of a velocity V, a characteristic length (full scale or model) 1, and the density of the fluid p, divided by the viscosity μ.

$$Re = \frac{VI\rho}{\mu} = \frac{VI}{v}$$

where v denotes the kinematic viscosity.

Ribbon Brake. Strap brake.

Rich Mixture. It is a fuel-air mixture in an internal-combustion engine which is having an excess of fuel over that required for correct combustion.

Rifling. Refers to the term which is used for the spiral grooves in the surface of the of a gun or rifle which are engaged by the driving bands of a projectile and cause rotation of the latter.

Rigidity Modulus. (G) G is given by $v = \dfrac{E}{2G} - 1$, where E denotes Young's modulus and v is Poisson's ratio.

Right-hand Engine. Refers to a horizontal engine which stands to the right of its flywheel as seen from the cylinder.

Right-hand Tools. Lathe side-tools having the cutting edge on the left, thereby cutting from right to left and putting the right-hand face on the workpiece.

Right-handed Engine. Refers to an aero-engine in which the propeller shaft is rotating clockwise when the observer is looking past the engine to the propeller.

Rim Wheel (rim pulley). Refers to a large pulley on the rim shaft of a cotton-spinning mule which is transmitting power to the roller which drives the spindles.

Rim Width (worm wheel). Refers to the maximum width of the rim in the direction of the axis of the worm wheel.

Rimer. Reamer.

Ring Gauge. Hardened-steel ring having an internal diameter of specified size used to check the diameters of finished cylindrical work. It has been manufactured to fine tolerances and used for go and no-go gauging.

Ring Lubricators. Flat metal rings, hanging from a shaft and dipping into a trough of oil, which rotate with the journal and carry oil to the bearing.

Ring Valve. Refers to a lift valve with a ring replacing the solid disc, the valve being guided by a central block fitting within the ring.

Rip-saw. A saw which has been designed for cutting timber along the grain with about one tooth per centimeter or three teeth per inch.

Rising and Falling Saw. Refers to a circular saw, whose spindle can be moved relative to the working table for cutting grooves of different depths.

Rising and Falling Spindle. Refers to a spindle of a circular saw which rises and falls by the operation of a worm and wheel.

Rising and Falling Table. A machine table which is used for drilling or other purposes which is raised for dropped to suit different requirements.

Rising Rod. Refers to a rod which is actuating the steam and exhaust valves in a Cornish engine through catches, sectors and weights.

Rivet. Refers to a metal pin having a circular shank and a head of various shapes. It is used for making a permanent joint between two mating parts having concentric holes. The rivet gets inserted through the aligned holes and made secure by hammering, pressing or expanding the tail of the rivet to form a second head on the far side of the assembly.

Riveting. Refers to the process of closing, clinching or hammering over the tail of a rivet to form a secure attachement of the rivet and the parts of the workpiece which it has been joining together.

Road Roller. A traction engine which is fitted with a large heavy roller for levelling road surfaces and the like. The term has been also used for smaller vehicles driven by diesel engines.

Roberval Balance. A weighing balance. Figure shows the principle of the balance. Two horizontal members of equal length are each pivoted at their centre-points, and two vertical members of equal length are joined to ends of the horizontal members by pivots. To the mid-point of each vertical member has been attached a horizontal arm on which weights are hung.

Robot. A reprogrammable device which is designed to manipulate and transport part, tools, or specialized manufacturing implements using preprogrammed motions to carry out specific manufacturing tasks.

A first generation robot is characterized by an absence of 'senses' (*e.g.*, touch, vision, smell) which would provide it with information about its surroundings. The second generation is having sensor-based control and considerable data-processing capability. The third generation has been provided with 'sensor' and decision-making capabilities.

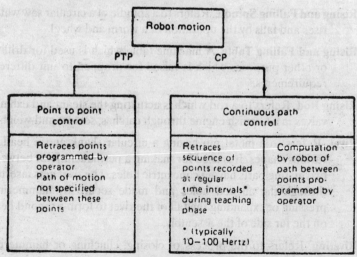

Robotic motion controls.

Robot Motion Control. There are two named types of robot motion control; Point-to-Point (PTP) and Continuous Path (CP) control.

Rock Drill. A reciprocating mechanism which is generally operated by steam, electricity or compressed air. The reciprocating rod is having a removable cutting head, which revolves slowly and is provided with feed gear.

Rocket-engine. An engine which is having within itself all the chemicals necessary for producing a propulsive jet, the chemicals being in the form of liquid propellants.

Rocket Motor. An engine which is having within itself all the chemicals, as solid propellant(s), necessary for producing a propulsive jet. (It has no moving parts in contrast to a rocket-engine).

Rocking Bar. A pivoted bar, connecting the winding stem in a key-less watch mechanism to the barrel for winding, or to the motion work for hand setting, and carrying the necessary intermediate wheel.

Rocket Frame. Refers to the frame which is carrying the gearwheels of a self-acting lathe to reverse the direction of the back shaft.

Rocking Shaft

    (*a*)   Refers to a shaft with a double-ended lever to actuate the solid valve in an indirect-acting slide valve type of steam-engine.

    (*b*)   Refers to a shaft or spindle with a to-and-fro motion only, such as a weigh shaft.

Rockwell Hardness Test. Refers to a commercial indentation test which uses a conical indenter for hard metals and a spherical indenter for soft metals. A small load of 10 kgs. The indentation has been directly recorded on a suitable dial after the further load has been removed.

Roll. Refers to the angular movement of a robot's end effector about an axis perpendicular to the wrist axis.

Rod

    (*a*)   A slender straight metal bar, frequently of circular cross-section, which may be part of either a structure or a mechanism.

    (*b*)   A bar.

Roll Feed. Nipping rollers which are used for feeding strip material to a power press. The rollers are timed in relation to the requirements of the press.

Roll Forming. Refers to the cold forming process which progressively forms strip metal into tubes or sections at high rolling speeds in a rolling mill.

Rollers. Refers to a cylinder which rolls or turns on its axis such as the rollers in a roller-bearing and in a roller chain.

Roller Bearing. Refers to a bearing on a shaft which is composed of a number of steel rollers and located by a cage between inner and outer steel races. Roller-bearings will carry heavier loads than ball-bearings. The rollers have been parallel for journal loads; if tapered, both journal and thrust loads can be carried.

Flexible-roller bearing. Hollow cylindrical bearings which are made by winding strip steel in the form of a helix.

Spherical roller-bearing. A roller-bearing having two rows of barrel-shaped rollers of opposite inclination working in a spherical outer race, thereby providing a measure of self-alignment.

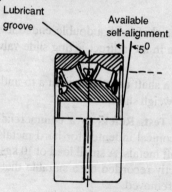

Spherical roller-bearing.

Roller Box. Refers to a cutting tool-holder which is used on capstan lathes an automatic laths. The box holds a cutting tool and two rollers positioned so that part of the reaction force from the cutting tool is taken by the rollers, thus preventing distortion of the work.

Roller Chain. See chain Figure.

Roller Conveyor. A power-driven or gravity-operated roller-track which is used for transporting packages or goods.

Roller Drive. A uni-directional drive. Figure shows three balls in the driving position where the outer member is going clockwise and one ball is in its free-wheeling position.

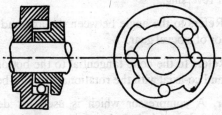

Roller drive.

Roller Mill. It consists of two rools which are mounted with horizontal axes and running in opposite directions for crushing and mixing operations.

Roller Steady. A support attached to the carriage of a lathe which is steadying the workpiece in the region of the cutting tool.

Rolling Curves (rolling circles). Templet curves which are used in striking out the shapes of gear-wheel teeth.

Rolling Locker. Refers to a traverse net lace machine having a double tier of carriages which swing in an are controlled by reciprocating rollers.

Rolling Mill. A set of rolls (Figure) having used for rolling metals into different shapes and sections like bars, billets, blooms, plates, rails, rods, slabs, strip.

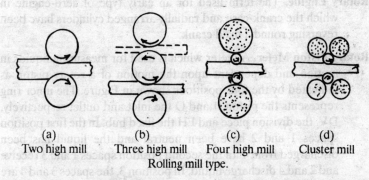

(a)	(b)	(c)	(d)
Two high mill	Three high mill	Four high mill	Cluster mill

Rolling mill type.

Rolls. Cylindrers of steel or cast-iron which are used in rolling mills. They are known as two-high or three-high according to the number of rolls one above the other. The former have been for rolling iron bar are plates, and the latter when the bar or plates have been passed backward through the upper rolls without the need for reversing.

Root Angle. Refers to the angle between the axis and the root cone generator of a bevel gear.

Root Cone. Refers to the cone tangential to the bottom of the tooth spaces and co-axial with the rotational axis of a bevel gear.

Roots Blower. A compressor which is used for delivering large volumes of air at relatively low pressure ratios, consisting of a pair of hour-glass shaped members rotating within a casing with but small clearance so that no valves are required.

Rope Wheel. A grooved pulley.

Rose Cutter. Refers to a small hollow milling cutter which is used by watch makers on a lathe for rapidly producing pivots, screws, etc.

Rose Reamer. A fluted broach having a conical serrated end and a sharp point (rose end).

Rotachute. Refers to a small rotocraft with blades that fold into a small compass, designed as a possible alternative to a parachute for use by one man. With the blades unfolded the descent of a man with rotachute has been in a steep glide, the rotachute behaving like an unpowered gyroplane.

Rotary Engine. The term used for an early type of aero-engine in which the crank-case and radially arranged cylinders have been reversing round a fixed crank.

Rotary Piston Meter. A meter which is used for measuring liquids in motion and dependent upon the motion of a rotary piston as indicated by the four positions shown in Figure. The inner ring represents the piston, 1 and O the inlet and outlet respectively, DV, the division piece and FH the fixed hub. In the first position spaces 1 and 2 have been neutral and the liquid has been discharged from 4. In the second position spaces 1 and 3 receive and 2 and 4 discharge liquid. In position 3, the spaces 3 and 4 are neutral, space 1 receives and space 2 discharges liquid. Finally in position 4, spaces 1 and 3 receive and 2 and 4 discharge liquid.

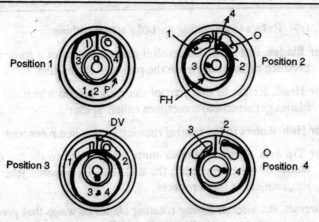

Position 1

Position 2

Position 3

Position 4

Rotary piston meter.

Rotary Planning. The term used for the work of making a surface with a vertical milling machine.

Rotary Pump (drum pump). Refers to a pump in which two specially shaped members rotate in contact. It has been suited to large deliveries at low pressures.

Rotary Squeezer. A machine which is consisting of two rotating eccentric cylinder, one external and the other internal. It is used for the consolidation of puddled ball.

Rotary Transfer Machine. Refers to a milling machine which is fit ed with milling cutters and multi-spindle drilling heads.

Rotary Valve (rotating valve). Refers to a cylinder valve which is rotating on cylindrical faces in a cylinder head and acts as a combined inlet and exhaust valve in a ported cylinder.

Rotative Engine. Refers to specifications which are used to distinguish the engine with a crank and flywheel from one in which the reciprocating movement has not been converted into circular motion.

Rotor

   (*a*)   Refers to the rotating part of a compressor or a turbine comprising the disc and its row of blades.

   (*b*)   Refers to a main rotor of a rotorcraft.

   (*c*)   Refers to a tail rotor of a rotocraft.

(*c*) Refers to a rotating part of a steam turbine.

Rotar Blades. Refers to the aerofoil-shaped blades of a rotor which produce thrust or increase the pressure of the fluid.

Rotor Head. Refers to that part of the rotor hub to which the rotor blades get attached; sometimes called 'spider'.

Rotor Hub. Refers to the central rotating system in a rotorcraft.

Rotor Tip Jets. Auxiliary power units which fitted to the tip of the blades of some rotorcraft; the units may be pressure jets, pulse jets, ramjets or small rockets.

Rotorcraft. An aircraft having rotating blades or wings that generate lift in flight, including helicopters, autogiros, gyroplanes, gyrodynes, convertiplanes and rotachutes.

The blades of a rotorcraft get hinged and the hinges are having descriptive names as follows: drag-hinge, feathering hing, flapping hinge. One type of helicopter has its blades hinged one-third of the way out on the blade.

Roughing Down (roughing). Refers to the removal of the bulk of the material in a machining operation.

Roughing Tool. Refers to a lathe, shaping machine or planer, tool for roughing cuts, having a round-nosed or obtuse-angled cutting edge.

Rotary Indexing Machine. Refers to a transfer machine in which tools or workpieces have been carried on a circular turntable which rotates intermittently, as required.

Router Cutter. A milling cutter which is driven through a chuck, the movement of which has been controlled by a guide pin running round a former, the former being a duplicate of the desired shape of the component which is being manufactured.

Routing Machine

(*a*) Refers to a machine with either (i) a fixed head in which the cutting head is having only a vertical movement on vee-slides, or (ii) a radial-arm type in which a similar head is mounted on the end of a long arm which makes it to move about a centre over an angular area of 350°. High-speed rotary cutters have been used to cut out shapes of varied profiles determined by formers, jigs, or numerical control.

(*b*) Refers to a machine with a revolving point which removes unwanted metal from printing plates.

Rubber Bond Grinding Wheel. Refers to grinding wheel in which the bonding material has been of rubber, which softens under the heat of grinding and acts both as a cushion for the abrasive grains and as a buff to polish out the grain marks.

Runner. Refers to the rotor (vaned member) of a water turbine.

Runners

(*a*) Devices which are used to assist sliding motion.

(*b*) Cylindrical sliding pieces supporting work in a pair of turns.

Running Centre Chuck. A driver chuck which is revolving with the lathe mandrel.

Running-in. Running a new engine, machine or mechanism under a light load and at moderate speed, to allow time for proper clearances to become established and friction-surfaces polished.

Running-out

(*a*) Refers to the working of a drill away from its centre.

(*b*) Refers to a piece of work/chucked eccentrically in a lathe.

(*c*) The deviation of a cutting tool from its required path.

Running Pulley

(*a*) Refers to the movable pulley in a snatch block.

(*b*) A gin pulley.

Run-out

(*a*) Refers to the total range of movement which is measured from a fixed point to a point on the surface of a gear or thread rotated about a fixed axis without movement along the axis.

Radial run-out. The run-out which is measured along a perpendicular to the axis of rotation.

Axial run-out (*wobble*). The run-out which is measured parallel to the axis of rotation, at a specified distance from the axis.

(*b*) A similar movement to (a) in other dimensional tests of gears.

(*c*) The quantity of additional-material to allow for all contingencies.

(*d*) The portion of a groove, serration, keyway, etc., where the full depth tapers away to nothing.

Rymer. Reamer.

S

G. Greek letter sigma. Symbol for stress; also f.

Shp. Shaft horse-power.

SI

- (*a*) International System of Units has been derived from the six basic units; meter, kilogram, second, ampere, kelvin and candela.
- (*b*) System International screw thread.

S/N Curve. See stress-number curve and Figure.

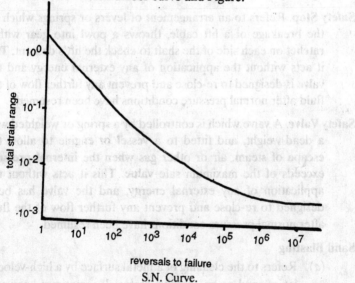

reversals to failure

S.N. Curve.

SV. Sluice or stop valve.

SWG. Standard Wire Gauge.

Saddle

- (*a*) Refers to that part of a machine that runs on slides.

(b)   Refers to that part of a lathe which runs the length of the lathe bed carrying the cross-slide, turret head or tailstock.

(c)   Refers to the sliding plate which carries the drill spindle and gearwheels of a radial drill.

Saddle Tank Engine. Refers to a steam locomotive having the water tank on the top and sides of the boiler.

Safe Fatigue Life. See fatigue life.

Safe Working Load. See factor of safety

Safety Factor. See factor of safety.

Safety Finger (guard pin). Refers to a pin or finger which butts against the edge of the safety roller of a clock if the escapement is jerked or, in the case of a watch, when the hands have been set back.

Safety Stop. Refers to an arrangement of levers or springs which on the breakage of a lift cable, throws a powl into gear with a ratchet on each side of the shaft to check the lift's descent. This it acts without the application of any external energy and the valve is designed to re-close and prevent any further flow of the fluid after normal pressure conditions have been resumed.

Safety Valve. A valve which is controlled by a spring or weighted with a dead-weight, and fitted to a vessel or engine to allow the escape of steam, air or other gas when the internal pressure exceeds of the maximum safe value. This it acts without the application of any external energy, and the valve has been designed to re-close and prevent any further flow of the fluid after normal pressure conditions have been resumed.

Sand Blasting

(a)   Refers to the cleaning of a metal surface by a high-velocity jet of sand or an abrasive material, sometimes known as shot-blasting.

(b)   Refers to the roughing of a metal surface by means similar to (a), prior to the application of a special finish.

Sand Pump. A centrifugal type of pump which is used for extracting wet sand out of caissons, etc.

Sanding. Feeding sand on wet or frozen rails in front of the driving wheels of a locomotive.

Sandslinger. A machine which is used for filling a mould with sand. It is delivered in wads at high speed by centrifugal force.

Saw. A tool with a serrated blade.

Saw Gumming. Refers to the grinding the roots or gullets of circular saw teeth by emery wheels, the sections of whose edges have been the counterparts of the tooth spaces.

Saw-sharpening Machines. Refers to machines fitted with abrasive wheels which automatically feed the saw tooth by tooth.

Saw Spindle. Refers to the spindle of a circular saw which is furnished with bearing necks and fast and loose pulleys. The saw is clamped between two washers and prevented from turning by a feather fitting into a slot cut in the saw.

Scavenging Stroke. Exhaust stroke.

Scavenging (or scavenger) Pump. Refers to an oil-suction pump which returns used oil from the crankcase of an engine to the oil tank, using the dry sump system of lubrication.

Scissors Jack. Four links in a parallelogram having a horizontal screw connecting one pair of opposite corners from this jack. The vertical corners get attached to the base and to the point where the load has been to be lifted. Rotation of the screw makes the horizontal corners together and at the same time separates the vertical pair thus lifting the load.

Sclerometer. An instrument which is used for measuring hardness. It consists of a diamond at one end of a lever attached to a vertical pillar. The diamond gets loaded and the pillar rotated to make a scratch of standard depth, the weight in grams to produce this depth giving a measure of the hardness.

Schrader Valve. An air check valve which is used in motor vehicle and bicycle tyres.

Scleroscope Hardness Test. Refers to the determination of the hardness of metals by measuring the rebound of a diamond-tipped hammer, weighing about two grams (one-twelfth of an once), when dropped from a given height.

Scotch Block. A block attached to rails to prevent the passage of rolling-stock.

Scotch Crank. A crank which is used on a direct acting pump. It consists of a square block pivoted on the over-hung crank pin and working in a slotted crosshead, which is carried by the common piston rod and ram.

Scotch Turbine. Refers to a development of a Barker's mill in which water gets admitted through a vertical supply pipe and flows outward through horizontal curved arms.

Scragged Springs. Figure shows the stresses in the cross-sections of the wires during the process of scragging.

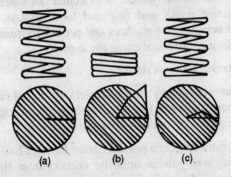

Resultant stress in scragged springs: (a) unstressed, (b) fully compressed, stress passes yield point, (c) resultant stress condition under no load.

Scragging Machine

    (*a*) A machine which is adapted for testing springs by impulsive loading.

    (*b*) A machine which is used for compressing springs to their solid length before use to increase their service life.

Scrap (capital scrap). Refers to scrap which is arising from absolescence and the discarding of manufactured goods.

    Circulating scrap. Refers to scrap which is arising in a steel works from ingot to saleable product.

    Process scrap. Refers to scrap which is arising from engineering and manufacturing operations to produce goods.

Scraper Ring. An auxiliary piston ring which is usually fitted on the skirt to remove surplus oil from the cylinder walls of a piston-engine and thus reduce oil consumption, the oil being led back to the sump through holes in the piston wall.

Scratch Hardness. Refers to the hardness of a mineral which is determined by the ability of one solid to scratch or be scratched by another.

Another measure of scratch hardness consists of drawing a diamond stylus under a known load across the surface to be examined, the hardness being measured by the width and depth of the resulting scratch.

Screw. A helix which is wound round a cylinder; an inclined plane which is wrapped round a cylinder.

Screw Barrel. A chain barrel having a continuous spiral groove to receive the edges of alternate links in the groove with the other links lying flat on the periphery.

Screw Compressor. Refers to a positive displacement rotary compressor. The gas is progressively compressed as it is forced between two mating helical screws in an axial direction.

Screw Conveyor. Worm conveyor.

Screw Cutting. Refers to the formation of screw threads on cylinders by taps and dies, by chasers, in a screw-cutting lathe, or by a traversing mandrel.

Screw-cutting Lathe. Refers to a lathe which is fitted with a lead screw and change wheels so that different rates of feed can be given to the slide rest relative to the rotation of the lathe mandrel. The lathe can cut screws or threads of different pitches.

Screw Gearing

   (a)   Gearing in which the teeth have not been parallel with the axes of the shaft.

   (b)   Refers to the transmission of power by means of a worm and worm wheel.

Screw Jack. A jack which consists of a vertical screw working in a nut and raised by the rotation of the nut by hand gear and a long lever, and provided with a ratchet.

Screw Machines. Machines which are used for turning and threading small screws, etc., from rod or bar fed through a hollow spindle.

Screw Micrometer. Micrometer guage.

Screw Pile. A pile with a wide screw or helix at the foot which gets rotated to drive it into the ground.

Screw Pitch. Refers to the distance between corresponding points on adjacent thread forms, measured parallel to and on the same side of the axis.

Also, refers to the reciprocal of the number of threads per inch or per centimeter.

Screw Pitch Guage. A small instrument having a number of guages, mounted like blades on a pocket knife, for ascertaining the number of threads per centimeter of a screw, (or threads per inch of a screw, usually ranging from 28 to 6).

Screw Plate. Tap plate A hardened-steel plate which is used for cutting small screws, in which there are a number of screwing dies of different sizes.

Screw Rate. Refers to the number of threads per inch.

Screw Thread. Refers to the ridge on the surface of a cylinder or cone which is produced by forming a continuous helical or spiral groove of uniform section and such that the distance between two corresponding points on its contour measured parallel to the axis has been proportional to their relative angular displacement about axis.

External threads are known as male threads and internal threads are known as female threads.

. *International screw thread*. Refers to a metric system in which the pitch of the thread has been related to the diameter the thread having a rounded root and a flat crest.

Metric screw thread. Refers to a standard screw-thread in which the diameter and pitch have been specified in millimetres.

Multi-start screw thread. A thread which is formed by a combination of two or more helical grooves equally spaced along the axis.

Parallel screw thread. A thread which is formed on the surface of a cylinder.

Single-start screw thread. A thread which is formed by a single continuous helical groove.

Taper screw thread. A thread which is formed on the surface of a cone.

Screw-threaded Gauge. Refers to an assemblage of thin steel plates whose edges have been notched to fit screws of different pitches for the purpose of identifying the pitch of a thread.

Screw Threads. Screw threads have been classified by their diameter, the number of threads N per inch or per centimeter and their form. Sometimes the pitch has been given, which has been the reciprocal of N.

A thread is right-handed if, when assembled with a stationary mating thread, it recedes from the observer when rotated clockwise. A left-handed thread recedes anti-clockwise.

Screw Tool. A tool which is used for cutting screws with a sectional shape the same as the interspace between two contiguous threads.

Screw-and-nut Steering Gear. Refers to a square-threaded screw on the lower end of the steering column of an automobile engages a nut provided with trunnions, which work in blocks sliding in a short slotted arm carried by the drop-arm spindle.

Screwing Machine. A lathe which is adapted for producing screws by means of dies.

Screwing Tackle. Appliances which are used for cutting screws apart from the aid of a lathe.

Scroll Chuck. A self-centring chuck with the jaws slotted to engage with a raised spiral on a plate which is rotated by a key to advance the jaws while maintaining their concentricity.

Scroll Gear. Refers to a form of variable gears in which the pitch surface has been in the form of a scroll, thus imparting a gradually increasing rate of motion to a shaft.

Scrubber Blocks. Refers to the blocks of cast iron which press on the thread of the wheels of high speed diesel-electric locomotives to condition the thread surface and improve the level of adhesion to the rails.

Scuffing Wear. Frictional wear due to a backward and forward rubbing between two surfaces.

Seat. Used to control the leakage of fluids into or out of parts of machines which may be stationary or moving.

Bellows seal. An annular ring of material bearing on the face of a rotating bearing on the face of a rotating part and mounted from bellows affixed to an adjacent stationary part. The bellows may enclose a rotating shaft or slide of varying length to keep it clean.

Cup seal. A disc-shaped with a single turned-up lip at its periphery used for sealing hydraulic and pneumatic pistons.

Sealing. Figure illustrates some types of sealing on shafts.

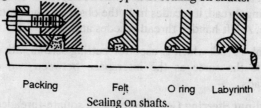

Packing   Felt   O ring   Labyrinth
Sealing on shafts.

Seaming Machine. A press which is used for forming and closing interlocking joints in the manufacture of sheet metal containers.

Second (S). Refers to the duration of 9192631770 periods of the radiation corresponding to the transition between the two hyperfine levels of the ground state of the caesium-133 atom.

Second Moment of Area (moment of inertia of an area). Refers to the sum of the products of each element of are, δA, multiplied by the square of its distance from an axis x, to give $\int dAx^2$ for a plane surface containing the X-axis. If the axis is perpendicular to the plane then it is known as the Polar Moment of Inertia.

Sector Gears. Refers to a form of toothed gearing with the wheels broken up into sectors of different curvatures, each pair of arcs transmitting different velocity ratios in an intermittent motion.

Segmental Gears (mutilated gears). Gears with teeth which have not been continuous around the periphery.

Seismograph. Refers to an instrument in which a heavy mass has been poised in such a way that a vibration of its support, together with the inertia of the mass, causes a relative motion of mass and support, that when amplified produces the record. The recording in older instruments has been by a stylus on a rotating drum and in more modern instruments an electro-magnetic current operates a mirror galvanometer to give a photographic trace. An observatory may have N-S and E-W horizontal instruments plus a vertical recorder.

Seizure of Seizing Up. Refers to the locking of two moving surface. such as in a bearing due to the partial welding together of the two surfaces, caused by insufficient lubrication of insufficient clearance between the two surfaces.

Self-aligning Ball-bearing. Refers to a ball-bearing with two rows of balls between an inner race and a spherical surface for an outer race, thereby allowing considerable shaft deviation from the normal.

Semi-rotary Pump. The alternate action of a semi-rotary pump is illustrated in Figure.

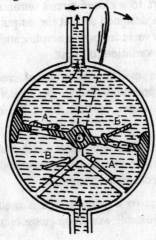

Semi-rotary pump.

Servo (servo mechanism). A device which is used for converting a small movement into one of greater amplitude or to exert a greater force.

Servo Control (in an aircraft). Refers to an additional mechanism which is devised to reinforce a pilot's effort by a relay.

Servo Mechanisms. Servo mechanisms have been of two types, with open loop or closed loop controlling systems. In the open-loop system there has been nothing in the mechanism to measure the result of the application and errors cannot be rectified. In the closed-loop system, the results of the operation have been fed back into the control circuits so that they can rectify errors. In guided missiles, the servo-loop has been closed by controlling impulses coming from the ground control and by error-sensing devices within the missile.

Servomotor

(*a*) Refers to a motor, linear or rotary, which receives the out-put from an amplifier element and drives the load, being the final control element in a servo mechanism. An example of a linear servomotor is hydraulic ram.

(*b*) Refers to a device for magnifying a small effort by using hydraulic means.

Servo System. Refers to a closed-circuit automatic control system, which has been designed so that the output follows closely the input; it usually includes power amplification and is capable of following rapid variations of input.

Servo Tab. Refers to a small hinged surface of an aeroplane control or flap, which is operated directly by a pilot to produce aero-dynamic forces which, in turn, move the control surface (or flap).

Servovalve. Refers to a hydraulic valve which makes a large flow of hydraulic fluid to be switched, by a very small initial force, from one part of a mechanism to another.

Setting Gauge. Refers to a gauge which is used for checking the setting of an adjustable workshop guage or an inspection gauge or a compressor.

Sewing Machine. A machine for sewing fabrics with a mechanism operating a needle bar, reciprocally vertically, having the thread for the pointed-eye needle supplied from a spool on the frame, plus usually a lock stitch mechanism.

Shaft. Refers to a spindle which is revolving in bearings and carrying pulleys, gear-wheels, etc., for the transmission of power. Shafts can be either solid or hollow, the latter giving an improved torque carrying capacity for a given shaft weight. They have circular cross sections where they fit into bearings and shoulders, collars, or steps of different diameter for the location of parts mounted on them. Shafts may run concentrically within one another as in a multiple spool gas turbine. Shafts may be having a taper shank to suit components with a matching taper bore, or vice versa, with a nut or tang giving positive location to the assembly, or the shaft may be threaded for rigid attachment of a component.

Shaft Angle. When the axes intersect as in bevel gears, the angle between these axes within which the pitch point lies: it has been equal to the sum of the pitch angles of the wheels and pinion.

Shaft Governer. A spring-loaded governor mounted on, and rotating about, the engine crankshaft for controlling the speed of small oil engines. It is sometimes housed in the flywheel. Shaft governors are also used to control the travel of the slide valves in steam-engines.

Shaft Straightener. Refers to a machine with three rollers adjustable for uniform pressure through which a shaft has been passed on to two larger rollers.

Shaft Turbine. A turbine which has been designed for transmitting power through a shaft, such as a free turbine to drive a helicopter rotor or propeller.

Shafting. Line shafting.

Shank

    (*a*)   Refers to the stem of a bolt, broach, cutting tools, drill, reamer, rivet, tap, which is held in the driving member.

    (*b*)   Also refers to the shaft of a tool connecting the head and the handle.

Shaping Machine (shaper). A reciprocating ram which carries a tool horizontally in guides or vertically in a clapper box and is driven by a quick return mechanism, for producing small flat surfaces, slots, etc.

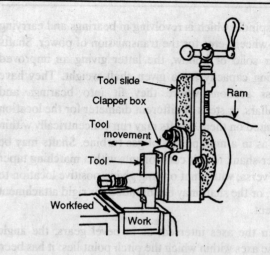

Shaping machine.

Shaving. Refers to a finishing operation in which a very small amount of metal has been removed from a blanked part.

Shear. Refers to a deformation in which parallel planes in a body remain parallel, but get displaced in a direction parallel to themselves.

Shear Face. Refers to the plane of the material which is subjected to shear from external forces.

Shear Force. Refers to the force that induces shear in body.

Shear Force Diagram. Refers to a graphical representation of the shear force acting on every part of a body which gives the values of shear in newtons.

Shear Modulus. With reference to the cross-section of a beam and to either principal axis it is the moment of inertia with respect to that axis divided by the distance from that axis of the most remote point on the cross-section.

Shear Stress. Refers to the shear force per unit area of the cross-section. Its value varies across the section being proportional to radius for torsion of a round bar or parabolic over the centre web of an I-beam.

Shearing. Refers to a method of cutting metals in sheet or plate from between two straight blades without the formation of chips.

Sheet. A term usually applied to rolled or extruded metal or plastic materials less than about 5mm (3/16 in) thick.

Sheet Mill. A rolling mill for the rolling of sheet metal.

Shims. Thin sheets of paper or slips or metal which are used as spacers to regulate the distances between objects, or as wedges to make parts of it.

Shimmy. Refers to the violent oscillation of a castoring wheel of an aeroplane's undercarriage, or of a vehicle, about its castor axis.

Shingling. Hammering or rolling puddled ball to convert it into bar or sheet iron.

Shock (motion). Refers to a motion of a transient nature which is caused by a very rapid change of displacement, velocity, acceleration, force or a very rapid change of temperature.

Shock Absorber (shock damper)

    (*a*)   Refers to an energy-absorbing device such as a frictional spring, rubber or hydraulic damper, in any part of a mechanism or moving vehicle.

    (*b*)   Refers to a bumper on a motor-vehicle.

Shop Traveller. Overhead travelling crane.

Shore Scleroscope. The term used for a machine in which a small indenter has been dropped from a specified height on to the specimen and the height or rebound, read on a graduated scale, is a measure of the dynamic hardness.

Shot Peening. Blasting the surface of a metal with small hard steel balls driven by an air blast to harden the surface layers.

Shoulder. Refers to that portion of a shaft or of a flanged structure where a sharp increase of diameter or other demension occurs.

Shrinking On. Fastening together two parts by heating the outer member until it expands sufficiently to pass over the inner, and on cooling grips it tightly; a process used for attaching steel tyres to locomotive wheels.

Shroud

    (*a*)   Circular webs which are used for stiffening the sides of the gear-teeth on a gear-wheel; called a 'shrouded wheel'.

(*b*)   Refers to a deflecting vane on one side of an inlet port of some piston-engines to promote swirl in the cylinder.

Shutter. A mechanical device in a camera or cinema camera for exposing photographic film for a known time and at will, using a roller blind with a slot, a rotating disc with a slot, an iris diaphragm or a circular eyelid device.

Side Frames. The main frames which are carrying the bearings for the shafts of engines, crarnes, pumps etc.

Side Valves. Poppet valves which are mounted in the side wall of an engine cylinder block.

Siemen (S). The unit of conductance $= 1/\Omega$.

Sight-feed Lubricator

(*a*)   A dip-feed libricator with a small glass tube supplying oil drops from a reservoir so that the drops can be seen.

(*b*)   A small glass tube filled with water so that oil from a pump rises in visible drops en route to the oil pipe.

Silencer

(*a*)   Refers to an expansion chamber which is attached to the exhaust pipe of an internal combustion engine to lessen the noise surrounding the engine.

(*b*)   Refers to an expanded portion in the exhaust system of any machine emitting a gas for dampening the sound transmitted from it.

Simple Steam Engine. Refers to a steam engine in which the steam expands from the intial to the exhaust pressure in a single stage.

Sine Bar. A tool which is used for the accurate setting out of angles by arranging to convert angular measurements to linear ones, thus

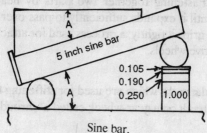

Sine bar.

depending on the accuracy of the bar length and the raising of one end by gauge blocks.

$$\text{Angle A} = \text{arc sin } \frac{1.000 + 0.250 + 0.190 + 0.105}{5} = 18°0.$$

Single-acting Engine. Refers to a reciprocating engine in which the working fluid acts on one side only of the piston, as in most piston-engines, in an early form of steam-engine and in steam-hammers.

Single-acting Piston. Refers to a piston in contact with the working fluid on one side only.

Single-acting Pump. A pump which is delivering liquid in alternate strokes only.

Single-cylinder Engine

(*a*)   An engine having only one steam (sylinder, plus a flywheel to carry it over dead centres.

(*b*)   An internal-combustion engine having only one cylinder, used for research purposes, for motor bicycles and for 'bubble' cars.

Single-cylinder Machine (yankee machine). Refers to a machine in which wet paper has been pressed on a polished heated cylinder and dried furing one revolution to impart a high glaze on one side. It is a common process in the production of high-class photographic prints.

Single-entry Compressors. See centrifugal compressor.

Single Gear. A gear which consists of one pinion and one wheel only in combination.

Single-plate Clutch. Figure shows a single-plate dry clutch which is operated through level E which pushes the collar and disc B forward with the assistance of springs D. It brings the clutch plate F along the splines, G, H to complete the contact between parts A and B via the clutch lining material K. The driving shaft is then transmitting its torque to shaft C.

Single Purchase. A lifting arrangement which consists of only a simple gear of pulley.

Sinking Pump. A pump which consists of two suction barrels with the pistons worked alternately with a bob lever, by hand or power, to clear water from foundations, etc.

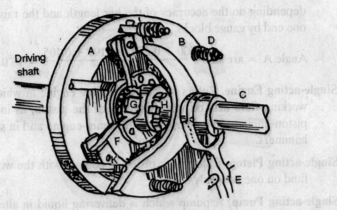

Single-plate clutch.

Sinusoidal. The trace of an alternating quantity plotted to a time base in the form of a sine wave; hence 'sinusoidally'.

Size. A generic term which denotes magnitude, especially geometrical magnitudes but also including weights, capacities, horse-powers, ratings etc.

Skew Bevels (skew bevel gears). Bevel gears whose axes have been not in the same plane.

Slabbing. Drawing down steel ingots into slabs using a stream-hammer.

Slabbing Machines. Large milling machines (planer type) which take wide and deep cuts off heavey work.

Slave Cylinder. In a hydraulic system it refers to any cylinder having a piston whose movement is controlled by movement of the piston in a master cylinder.

Sleeve. Refers to a tubular piece, usually machined externally and internally, into which a rod, another tube or piston has been inserted.

Sleeve Valve. Refers to a thin steel sleeve which is fitted within the cylinder of a petrol or oil engine. Rotation of the engine's crankshaft imparts an osciallatory axial motion of the sleeve which brings the sleever-port (hole) in line with an inlet or an exhaust port in the cylinder wall at the appropriate time.

Sleeve Wheel. A wheel having a long hollow boss which fits over and slides along a shaft.

Slenderness Ratio. For a strut in compression this is the ratio of its length to its radius of gyration.

Slewing Gear. Refers to the gear for slewing a travelling crane. It consists of friction cones which actuate a vertical shaft; this last revolves a pinion in a curb ring on the crane's truck.

Slide

    (*a*)   Refers to a piece of mechanism which moves in a linear direction over a flat or curved smooth face between guides.

    (*b*)   Refers to the operating piece on the outside of a repeater.

Slide Blocks (guide block, slipper blocks). Blocks which are attached to an engine or pump crosshead and moving between guide bars, which ensure that the path of the piston rod has been truly rectilinear.

Slide Valve

    (*a*)   Refers to a valve which slides in contrast to one that rotates or lifts, referring usually to the valves of steam-and gas-engines.

    (*b*)   Refers to a steam-engine inlet and exhaust valve shaped like a rectangular lid, alternately admitting steam to the cylinder and connecting the ports so exhaust through the valve cavity.

Sliding. Refers to the motion of a slide rest on its bed.

Sliding Friction. See friction.

Sliding-head Automatic Lathe. Refers to a lathe with a sliding head-stock the cutting tools being held radially in a tool bracket fixed to the lathe bed, each tool being brought into operation in the correct preset order by means of cams and levers.

Sliding Mesh Gearbox. Refers to a gear-box in which the gear gets changed by sliding one pair of wheels out of engagement an sliding another pair in.

Slipper Brake. An electromechanical brake which is acting directly on the rails of tramway.

Slipper Guides. Guide bars.

Slipper-piston. A light piston with its skirt cut away between the thrust faces, thus saving weight and reducing friction.

Slipper. Retarder.

Slot. Refers to a long groove in a machine part into which another part slides. A groove into which a mating member can get inserted for the transmission of torque, such as a screwdriver slot.

Slot Machine. Refers to a retail dispensing or amusement device in which a mechanism operates upon the insertion of a coin.

Slotting Machine (vertical shaping machine) (slotter). A machine tool which is having a ram with a vertical motion balanced by a counterweight and cutting on the downstroke towards the table. It has been similar to a shaping machine.

Slotting Tools. Tools used for cutting keyways, etc; in a slotting machine, having a narrow edge and deep, stiff section, with top and side clearance but little rake.

Slow Gear

    (*a*)   Refers to the double or multiple purchase gear of cranes and hoisting machinery.

    (*b*)   Refers to the back gear of lathes, drilling and other machines giving an increase of power with slower movement.

Sluice Gate. A barrier plate free to slide vertically across a water channel or an opening in a lock gate.

Snift Valves. Valves which open for the passage of air but close if liquids attempt to pass.

Snout Boring Machine. A boring machine with the cutters carried on an unsupported extension of the spindle. It is used for boring engine cylinders.

Snubber (snub). A device for limiting the movement of a system and not coming into action until the displacement of the system reaches a predetermined amount; for example, the checking of a cable or rope suddenly when running out.

Socket. A hallow for something to fit into, or stand firm, or revolve in. A bayonet fitting has been a type of socket.

Solar Gear. An epicyclic gear having a rotating annulus, a fixed sun wheel, a rotating planet carrier and planet wheels rotating about their own spindles.

Soldering. Refers to the hot bonding of metal parts, using a thin film of law melting point alloy (solder).

Sole Plate (sole). The bedplate of a marine engine.

Solid Coupling. A coupling forged in a solid piece with its shaft, as in a propeller shaft.

Solid Injection (airless injection). Injection of liquid fuel, by a high pressure pump, into the cylinder of an oil engine using the diesel cycle.

Spark Erosion. Refers to an electro-chemical machining process in which an electrode, often of complex cross-section is maintained very close to the work-piece within a dielectric liquid repelling particles from the work-piece and reproducing the shape of the electrode.

Sparking-plug (spark). A plug which is screwed into the cylinder head of a petrol engine to get ignition of the fuel-air mixture in the cylinder by a spark discharge between the insulated central electrode and one or more earthed points.

Specific Fuel Consumption (internal-combustion engine). Refers to the mass of fuel consumed by an engine per unit of delivered energy (kg/MJ or horse-power/hour).

Specific Speed

(*a*) Refers to the speed at which a water turbine will run to produce one horse-power under a head of one foot.

(*b*) Refers to the speed at which a pump will deliver one gallon of water per minute under a head of one foot.

Speed. Refers to the ratio of the distance, in a straight line or in a continuous curve, covered by a moving body to the time taken. Rotational speed may be measured by the number of complete revolutions in a given time. Speed is a scalar.

Speed Indicator. See speedometer; tachometer.

Speed Governing. See governors.

Speed of Rotation. See speed.

Speedometer. A tachometer which is fitted to the gear-box or propeller shaft of a motor other road vehicle and usually graduated to indicate the speed in miles (and/or km) per hour. The centrifugal type has been specially described (see speedometer, centrifugal) and reference has been made there to other mechanical types; in addition, there are electrical air- vane types and magnetic 'drag cup' types.

Speedometer, Centrifugal. Mechanical speedometers usually employ one of three principles, namely, centrifugal, chronometric or magnetic. Figure shows the first, the second has a clockwork movement to determine the quotient of distance divided by time and the third depends on a rotating magnetic field in an air gap. In the illustration the weights have been thrown outwards on the governor principle, the arm of which has been hinged to a rotating arbor. The arm has been linked to a sliding sleeve on the arbor and throught this the position of the weights controls the instrument pointer.

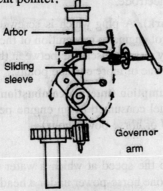

Arbor

Sliding sleeve

Governor arm

Centrifuga speedometer.

Speeds

    (*a*)   The number of steps of a cone, driving pulley or drum.

    (*b*)   The rotation speeds of a lathe chuck or machine tool.

Spherometer. An instrument with a central micrometer screw for measuring the height of convex or concave surfaces above or below a zero mark.

Spider

    (*a*)   Cathead.

(*b*)   A spoked central member, fitting on a propeller shaft which transmits the drive to the blades mounted one on each spoke. The pitch of the blade can be altered by rotation of the spoke.

(*c*)   The spoked central member of a universal joint.

Spigot Bearing. A bearing which carries two shafts in line while allowing them to turn independently.

Spindle. A slender metal rod or pin on which something turns.

Spindles. See breaking pieces.

Spinner. A streamline fairing, enclosing the hub or boss of a propeller, fitted coaxially and rotating with the propeller.

Spinning Wheel. A household instrument which is used for spinning yarn or thread with a flywheel driven by crank of treadle.

Spiral Bevel Crown Gear. A gear whose pitch curves have been inclined to the pitch element at the spiral angle and are usually circular.

Spiral Gear. Crossed helical gear.

Spiral Winged Valve. A lift valve having wings arranged as sections of a spiral of very long pitch so that the valve turns round on its seating to a slight degree at each lift thus ensuring uniform wear.

Spiroid Gears. A type of worm gear which has been conical in shape with the mating member a face-type gear. A large number of teeth have been in simultaneous contact, very high ratios can be accomplished in a single step, and these gears can be manufactured by standard hobbing machines.

Splashers (guard straps). Refers to the sheet iron strips which arch over the tops of locomotive wheels as a protection against injury to the drivers.

Splines. Relatively narrow keys integral with a shaft and resembling long gear-teeth.

   External splines. Splines produced by milling longitudinal grooves in a shaft.

   Internal splines. Grooves which are formed in a hole into which a splined shaft fits.

Splined shaft.

Split Bearing (divided bearing). Refers to a shaft bearing with split housing and the bearing bush (or brasses) clamped between the two parts.

Split Gear (Split-and-Sprung Gear). Refers to a fixed and a loose gear-wheel which is mounted alongside on the same shaft and coupled together by springs, pre-loaded so that backlash gets eliminated when both gears mesh with a common rack, gear-wheel or pinion. Sometimes known as 'anti-backlash gear'.

Split Ring. A ring divided either diagonally or with a lapped joint, such as a piston ring.

Split Wheel. A split cog-wheel, split for a use similar to a split pulley.

Spoke

    (*a*)   A radial bar of a wheel or of a steering-wheel of a vehicle.

    (*b*)   A rung of a ladder.

    (*c*)   An arm of a running wheel of a locomotive or rolling stock.

Spoke Machine. A copying machine or lathe with two sets of centres having a template in the one and the piece of work in the other, a roller moving on the template and a cutter following, being set in the same slide rest.

Spool. A gas-turbine engine which is having two compressors driven separately by two turbines is called a two-spool jet-engine; hence the terms single-spool compressor and two-spool compressor. The highest practicable compression ratio of a single-shaft engine has been about 7 to 1, but a two-spool engine may be more than 9 to 1.

Spool Valve. A piston valve.

Sprag Clutch. See clutch.

Spreadboard. Refers to an endless belt upon which handfuls of sorted flax fibre are laid for conversion into sliver.

Spreader. In a double disc gate valve the 'upper spreader' has been the component attached to, or engaging, the actual thread of the stem and the 'lower spreader' it its complementary component. Together in conjunction with the stop in the body, they constitute the spreading mechanism to force the discs apart against the body seats when the valve is closed.

Spring Balance, Circular. A weighing machine dependent on the extension of helical wire spring the extension being proportional to the load, provided the wire is not stretched beyond its elastic limit. In circular spring balances the springs have been used in one or more pairs, as indicated in Figure. The extension of the springs moves the rack and thus actuates the pinion carrying the pointer. A small spring presses the rack into contact with the pinion.

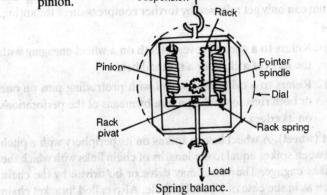

Spring balance.

Spring Box. A hollow box having a helical spring which is thus constrained to exert a force along one axis only.

Spring-loaded Governor. Refers to an engine governor in which the rotating masses, which move outwards under centrifugal force, are controlled by a spring.

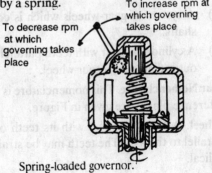

Spring-loaded governor.

Spring Rate (spring constant). Refers to the ratio of load to deflection for a spring. For many types of metallic springs this rate has been approximately a constant.

Spring Ring. A metallic ring which is cut diagonally and bigger than the bore to render the close fitting of a piston when spring into an engine cylinder. A Ramsbottom-ring is a spiral of two or three coils or separate cut rings pressed into grooves in the periphery of a solid piston.

Spring-set Brake. Refers to a double block brake in which the brake shoes have been held in contact with the drum by a spring. The drum can only get released by further compression of the spring.

Sprocket

(a) Refers to a each of several teeth on a wheel engaging with the links of a chain or a toothed belt.

(b) Refers to a cylindrical wheel with protruding pins on one or both rims for pulling a film by means of the perforations on its edges.

Sprocket (wheel). A wheel having spikes on its periphery with a pitch between spikes equal to the lengths of chain links with which the spikes engage. The wheel may drive or be driven by the chain such as in the case of a pedal bicycle. Also called 'bracket chain wheel'.

Sprocket Chain. Pitch chain.

Spur Gear (spur gearing)

(a) Gearing which is composed of combinations of spur wheels.

(b) A system of gear-wheels which is connecting two parallel shafts.

(c) A cylindrical gear with teeth parallel to the axis and on the outer diameter; a spur wheel.

Spur-gear Nomenclature. This nomenclature is shown in Figures. The reference planes are given in Figure.

Spur Wheel. A toothed wheel with its teeth on the outer diameter parallel to the axis. The teeth may be straight, helical or double helical.

Square-nose Tool. Refers to a finishing tool for turning with its cutting edge at right-angles to the edges of the shank.

Squeezing Machine. A machine in which bars or rails have been bent or straightened.

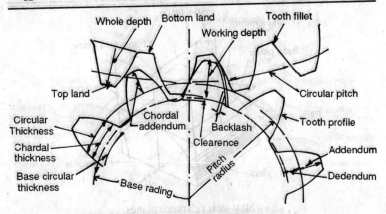

Spur- helical,- and bevel-gear nomenclature.

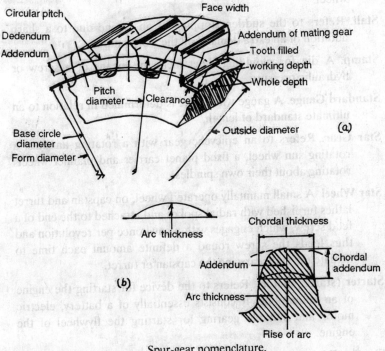

(a)

(b)

Spur-gear nomenclature.

Staggered. A term to describe (i) A linear distance between the leading edges of a series of parts in an assembly. (ii) The

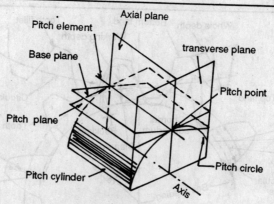

Spur-gear reference planes.

alternate inclination in opposite directions of the spokes or a wheel.

Stall. Refers to the sudden stopping of an engine due to a sharp addition of an extra demand on the work output.

Stamp. A die for moulding sheet metal by means of a screw or hydraulic pressure.

Standard Gauge. A gauge whose size is determined in relation to an ultimate standard of length.

Star Gear. Refers to an epicyclic gear with a rotating annulus, a rotating sun wheel, a fixed planet carrier and plannet wheels rotating about their own spindles.

Star Wheel. A small manually operated wheel, on capstan and turret lathes furnished with radial spokes and attached to the end of a feed screw, which engages with a pawl once per revolution and thus feeds the screw round a definite amount each time to actuate the slide carrying the capsian or turret.

Starter (starter motor). Refers to the device for starting the engine of an automobile. It consists essentially of a battery, electric motor and suitable gearing for starting the flywheel of the engine.

Starting Engine

(*a*) A small engine which is used for starting large diesel and steam-engines.

    (*b*) A small auxiliary gas-turbine engine which is carried by some aircraft to provide ground power supplies; including high-pressure air for starting the main engines.

Starting Lever. The lever which is attached to a starting slide valve.

Starting Valve. A valve which is used for admitting steam from the boiler to the cylinders of a steam-engine.

Static Stiffness. (See also stiffness). Refers to the change in deflection with change in dead load.

Stator. A row of radially disposed stator blades.

Stator Blades. Fixed blades of aerofoil shape mounted on the casing of an axial compressor to direct the airflow between the several rotary stages. The first row in many engines has been adjustable. The same term has been used for the stationary blades performing a similar function between the rows of turbine rotor blades.

Stator Blades (exhaust). An assembly of stator blades situated behind the axial compressor or turbine discharge to remove residual swirl from the exhaust gases and usually in sections to allow for thermal expansion.

Stave

    (*a*) The projecting end by which a crankshaft has been held during forging.

    (*b*) The shaft upon which a large number of healds have been mounted.

Steam Chest. The chamber to which the steam pipe is connected and in which the slide valve works.

Steam Crane. See crane.

Steam Dome. See dome.

Steam Donkey Pump. A small pump which is used for supplying steam boilers with feed water, having the piston rod of the steam cylinder and the ram of the pump in line, and a flywheel to carry the momentum through the dead-centres.

Steam-engine Cycle. Steam has been admitted to a cylinder on one side of a steamtight fitting piston. The piston has been connected by a piston rod to a crosshead which slides between guides and the crosshead moves the crank through a connecting-rod. After expansion in the cylinder the exhaust steam either passes to a condenser or through two or more cylinders of increasing size. The admission and exhaustion of the steam from each cylinder has been controlled by a slide valve.

Steam Engines. Engines in which the motive power has been supplied by the elasticity and expansion of rapid condensation of steam.

Steam Gauge. See Bourdon gauge.

Steam Hammer. A heavy hammer used for forging which slides between vertical guides. Steam is used to raise and sometimes to accelerate the descent of the hammer which otherwise falls freely under gravity.

Steam Line. Refers to the upper approximately straight portion of an indicator diagram which records the entrance of the steam into an engine cylinder.

Steam Locomotive. A steam-engine and boiler integrally which is mounted on a frame for service on a railway.

Steam Ports. Refers to passages supplying and exhausting steam from the valve face to the cylinder of a steam-engine.

Steam Pump. A force pump which is driven by a steam-engine.

Steam Reversing Gear (steam reversing cylinder). A power reversing gear of a steam locomotive which is operated by a reversing lever to admit steam to an auxiliary cylinder which operates the reversing links of the valve gear.

Steam Tables. Tables of figures providing the properties of steam over a pressure range.

Steam Trap. An automatic device which is used for ejecting condensed steam pipes, etc., without permitting the escape of steam.

Steam Turbine. Refers to a turbine in which the steam does work by expanding, thus creating kinetic energy, as the steam passes over the blades of the turbine disc (or drum).

Steam Valve. A valve which is used for regulating the supply of steam.

Steering Head. Refers to the gear for steering a ship.

Steering Gear. The two geared members transmitting motion from the steering-wheel of an automobile to the stub axles through the drop arm, drag link, steering-arms and track-rods.

Steering Lock. Refers to the maximum amount that the steered wheels of a vehicle can turn from side to side.

Steering Wheel. Refers to the spoked handwheel at the top of the steering column of an automobile which actuates the steering gear.

Steradian (Sr). Refers to the solid angle which is subtended at the centre of a sphere by a cap with an area equal to the radius squared.

Stern Tube. Refers to the tube which carries the bearings of a ship's propeller shaft.

Stick-slip Motion. Refers to the motion of sliding material when the force to start a surface sliding is greater than that to continue it moving. That is when the stiction is greater than the friction.

Stiffness. Refers to the resorting force per unit displacement. The reciprocal of compliance.

Stiffness ratio. Refers to the ratio of stiffness along two defined principal axes in isolators and in rigid bodies with resilient supports.

Stirling Engine. Refers to reciprocation external combustion engine having two pistons working in one cylinder, a working piston and a displacer piston.

Stock. Refers to the principal part or body of a tool or instrument to which the working part gets attached.

Stocking Cutter. A milling gear-cutter which is used for roughing out the material between gear-teeth, to be followed by a finishing.

Stockless Anchor. A anchor having no cross-piece on the shank and the arms pivoted so that both can engage at the same time.

Stocks and Dies. Dies and their holders for cutting screws.

Stokes (St). It is a unit of kinematic viscosity, usually quoted in centistokes (cSt); $1 \text{ cSt} = 10^{-6} \text{ m}^2/\text{s}$.

Stop. A metal pin or block to stop or reverse the action of a machine or mechanism.

Stop Cock

(*a*) A cock.

(*b*) A plug.

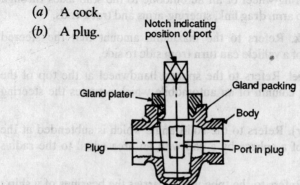

Common stop cock.

Stop Motion. A lever arrangement which is used for disengáging a machine or mechanism without stopping it.

Stop Valve

(*a*) Refers to a valve which, when shut, prevents the flow of liquid or gas.

(*b*) The main steam valve on a boiler for controlling the steam supply and isolating the boiler from the main steam pipe.

Straight Bevel Crown Gear. Refers to a gear whose pitch curves have been straight lines, intersecting the apex. The spiral angle at any cone distance is zero.

Straight-flute Drill. A conical pointed drill with straight longitudinal flutes in the shank.

Straight-line Lever Escapement. See Swiss lever under escapements.

Straight Link. A slot link with the reversal movement divided between the link and the blocks.

Strain. When a material gets distorted by external or internal forces acting on it, it is said to be strained. Strain may be defined as the ratio: Deflection Dimension of material over which the deflection has been measured.

$$e = \frac{\Delta l}{l}$$

It has no units.

Compressive or tensile strain (E or ε). This strain may be defined as the ratio · Contraction or elongation/Original length.

Shear strain (ϕ) (*detrusion*). Shear strain may be defined as the ratio: Deflection in direction of shear force/Distance between the pair of shear forces.

Volumetric strain (ϕ) Volumetric strain may be defined as : Change of volume/Original volume.

Strain Gauge. A device which is used for converting mechanical strain into a measurable electrical signal. It consists of a very thin metal wire, metal foil or semiconductor filament bonded on to a backing sheet by which it can be attached to a body before it is put under strain. As the body gets strained the electrical conductor gets correspondingly strained. This alters its electrical properties and with prior calibration the value of the strain is obtained. A Wheatstone Bridge circuit has been used for static analysis and temperature compensated dynamic analysis, whilst a potentiometer circuit can be used for direct readout, dynamic analysis when temperature compensation is not needed.

Strain-bardening. Increase in resistance to deformation due to prior deformation.

Strap Brake. A brake which consists of a strap or belt, sometimes fitted with shoes, enveloping a flywheel.

Stress. Stress may be defined as the force per unit area. There have been three kinds; tensile, compressive and shear. Bending produces both tensile and compressive stresses while torsion produce's a shear stress. An alternating stress exists when a force keeps changing sign. An alternating stress can be superimposed on a steady stress and the resultant combination may be of alternating sign or the steady stress may so dominate that the resultant has been of one sign but, of course varying magnitude.

Stress Concentration. Refers to any interruption to the smooth flow of stress and strain in a component, like a hole, corner, groove, notch or crack, introduces a concentration of stress, which has to be accounted for in calculating the strength of the component especially under fatigue conditions.

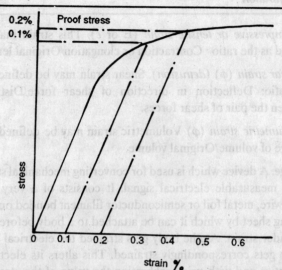

Stress/strain curve showing derivation of proof stress.

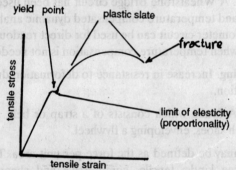

Stress/strain relationship for a material having yield point.

Stress Concentration Factor. If a stress concentration exists a further factor, for the particular type of stress raiser, has been added to any factors considered necessary under ordinary stressing conditions. The stress concentration factor has been the ratio of the true maximum stress to the stress calcuated by the standard mechanics formula using the net cross-section properties.

Stress Corrosion. Certain metallic alloys have been susceptible to premature failure under sustained tensile stress in a corrosive environment. This stress corrosion takes place when the material is kept in a corrosive environment under tension of sufficient magnitude and duration to allow the initiation and growth of cracks. Failure then takes place at a stress which is lower than that which the material would normally be expected to withstand.

Stress Ratio. Refers to the ratio of maximum to minimum stress in fatigue, taking signs into account.

Stress-number Curve (S-N curve). Refers to a curve which is obtained from fatigue tests in which a series of specimens of a given material have been subjected to different ranges of stress. The range of stress has been plotted against the number of cycles required to produce failure. In steel and many other metals there has been a limiting range of stress below which even an infinite number of cycles will not produce failure. This is often known as the fatigue limit.

Stress-strain Diagram. A plot of stress versus strain, usually but not necessarily tensile.

Stress-strain Relationship. Refers to the effect of increasing the stress on a material and its corresponding increase in strain which will have a unique relationship for each material. Shows a typical stress-strain curve for a mild steel. For stresses up to the elastic limit the material will return to its original length upon removal of the stress, and over this length of the curve the ratio of stress over strain is a constant known as the Young's Modulus of the material, and the material is obeying Hooke's Law. Under the influence of loads inducing stresses above the yield point the material has been no longer elastic and after passing through a plastic state will eventually fracture.

Striking Gear. Refers to the lever, fork and essential fitting for shifting belts on and off the pulleys of rotating shafts.

Stripper and Worker. Refers to a pair of wire-covered rollers which are part of a roller carding engine.

Stripping. Damaging the teeth of wheels due to too sudden starting and stopping.

Stroboscope. A device which is used for measuring the rotational speed of a rapidly rotating object appears to rotate at a speed equal to the actual speed difference of the device and the object and appears fixed when the two are synchronized. The object has been usually observed (*a*) through a slot in a rotating disc or drum, or (*b*) with the aid of flashing light of pre-fixed frequency.

The device can be used for the detection of wear, distortion or chatter of moving parts and mechanisms.

Stroke

(a) Refers to the total length of the movement of a piston in the cylinder of a piston-engine.

(b) Refers to the length of travel of a reciprocating part of an engine or mechanism.

Stub-tooth gear. A gear with robust teeth of short height as used in the manufacture of automobile gears.

Stuffing Box. Refers to an annular space through which a machinery part moves, and in which packing has been composed by a gland to make a pressure-tight joint such as the rod of a pump or the stem of a valve.

Suction Box. Refers to the lower chamber of a suction pump into which the liquid has been drawn by the upward stroke of the piston.

Suction Dredger (sand-pump dredger). A dredger which uses the suction of a centrifugal pump to bring material up a long pipe to discharge it into a barge.

Suction Pump. Refers to a pump which depends on atmospheric pressure for its action and which, in practice, will not lift water from a depth greater than about 7.6 m (25 ft). The water gets drawn into the barrel by suction. The descent of the piston allows the water to pass through a bucket valve into the upper part of the pump from which it gets lifted out and delivered on the upward stroke.

Summing Gear. Refers to a differential gear which adds or subtracts the motions of two members.

Sump

(a) Refers to a lowest part of the crankcase of an engine which acts as the reservoir for the lubricant.

(b) Refers to a similar location in any machine which requires lubricating.

Sun and Planet Wheels. Refers to a gear wheel (sun wheel) around which one or more planet wheels or planetary pinions rotate in mesh.

Supercharger. Refers to an axial flow or centrifugal compressor which supplies air, or a combustible, mixture, to a piston-engine at a pressure greater than atmospheric and gets driven either directly by the engine or by gas turbine motivated by the exhaust gases. The latter is termed as an 'exhaust-driven supercharger' or a 'turbo-supercharger'.

Supercharging

(a) Refers to the maintenance of ground-level pressure in the inlet pipe of an aero-engine up to the rated altitude by a supercharger.

(b) Boosting.

Superfinishing. A process which involves short strokes at a very rapid rate with a lighter pressure than in honing and lapping and with copious amounts of coolant and lubricant. It could be applied to both cylindrical and flat surfaces and surfaces finish is usually in the range 25-100 nm (1×10^{-9} to 4×10^{-9} in) 10 nm (4×10^{-10} in) can be obtained.

Superheated Steam. Refers to steam which is heated at constant pressure to a temperature above that dure to saturation and out of contact with water from which it was formed.

Super Miser. Refers to a combination of air preheater and economizer for boilers.

Surface Chuck. Face chuck.

Surface Condenser. A condenser in which cooling water gets circulated in tubes to condense the steam and a vacuum is maintained by an air-pump.

Surface Grinding Machine. A machine which is used for finishing flat surfaces with a high-speed abrasive wheel mounted above a reciprocating, or rotating, work table on which the work is held often by a magnetic chuck.

Surface Meter. An instrument which is used for measuring the texture of surfaces, using a stylus whose up and down motion is magnified up to 100,000 times to produce a graph of cross-section of the surface and a number representing the centre-line average height.

Surface Plate (planometer). A rigid, accurately flat, cast iron plate which is used for testing the flatness of other surfaces and to provide a plane datum surface for marking off work for machinery.

Surface Texture. The term used for the appearance and characteristics of a metal surface after machining.

Surging

(*a*) Refers to a severe fluctuation or abrupt decrease in the delivery pressure of a centrifugal supercharger or of a compressor.

(*b*) Refers to the coincidence of a harmonic of a cam's lift curve with its controlling valve spring's natural frequency of vibration, leading to irregular action.

Suspension. Refers to that portion of a mechanism which is designed to damp vibration and reduce the effect of external shock loads on the major portion of the assembly.

Suspension Links (vibrating links). Refers to two parallel flat rods which lift and lower the slot links of a steam-engine for reversal. One end of the pair has been loosely attached to the tail of the slot link and the other is attached to a short lever keyed on the weight shaft.

Swashplate (wabbling disc). A circular plate which is mounted obliquely on a shaft, as a substitute for a crank mechanism.

Sweat Cooling. Refers to the cooling of a component of an engine or mechanism by evaluating fluid through a porous surface layer, like in rocket-engines and gas turbine blades.

Swiss Machine (schiffle or shuttle machine). An embroidery machine having the shuttles placed diagonally.

Swivel-head Lathe. A special lathe which is used for boring and turning tapered objects having the mandrel headstock mounted and pivotable on a base plate.

Synchromesh Gear. Refers to a gear in which the driving and driven members have been automatically synchronized by small cone clutches before engagement.

Synchronizing Gear. A gear to synchronize the firing mechanism of a gun with the rotation of the airscrew so that the bullets do not meet the blades.

T

θ. Greek letter theta. The symbol for angle of twist.

TS. Tensile Strength ultimate.

Table. Refers to the horizontal portion of a machine on which the work is kept for planning and other operations.

Tachometer. Refers to an instrument which is used for indicating the revolutions per minute of a revolving shaft, operated either by a spring-controlled ring pendulum or by sprin-loaded governors, or by magnetic means. When registering the revolutions of a revolving shaft in cotton spinning it is known as an 'indicator'.

Tail (auxiliary) Rotor. A small rotor which is mounted at the tail of a helicopter on a horizontal axis to provide sideways thrust to contract the torque of a single main rotor, and to give directional control; also called 'auxiliary rotor' anti-torque rotor'.

Tail Stop Screw. The back screw of the headstock of a back-geared lathe.

Tailshaft. The shaft driving a marine propeller.

Tailstock (poppet). The movable head of a lathe which supports the end of the work remote from the driving headstock.

Tandem Engine. An engine having the cylinders arranged axially, or end to end, with a common piston rod.

Tangent Screw (tangent wheel). A worm screw which is used for making fine adjustment in the setting of an instrument about the worm wheel axis to correct its line of sight or to adjust a vernier.

Tank Engine. A locomotive having a tank for carrying its own water, thus dispensing with a tender. The tanks are either on the side (side tank) or on the top of the boiler (saddle tank).

Tape Condenser. A mechanism which controls leather tapes and rubbing rollers for converting the web of fibres from the doffer of a carding engine into a number of silvers.

Tape-controlled Machine. Any machine whose operation has been in part, or wholly, controlled by a punched-tape system. Usually the operation of feed and speed of a cutting tool or the workpiece position has been controlled by electronically reading instructions from the tape. Optical and pneumatic reading coupled with mechanical, pneumatic or hydraulic actuation is sometimes used.

Taper Pin. Refers to a pin locking member between two mating parts. Figure shows two applications for a taper pin.

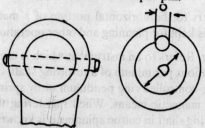

Taper pin application.

Taper Roller-bearing. A roller-bearing with tapered rollers to take end thrust and radial loads using internally and externally coned races.

Tapper Shank. A shank which is tapered in its length and circular in section as in some drills and reamers.

Tapper (tappet valve). A sliding member working in a guide to operate a push-rod or valve system. The tappet's motion has been controlled by a can and the sliding motion eliminates side thrust and distinguishes it from the motion of a pawl or screw. The tappet and cam convert a rotating into a reciprocating motion. In weaving, a tappet mechanism is called a 'wiper'.

Figure shows a tappet valve in a side-valve engine. A being the mushroom head.

Teaching. Refers to a method of programming a robot by driving it, or manually leading it, through the required sequence of motions. The recorded program is then used for automatic control of the robot. In the case of manual leading, recording has been usually done automatically at chosen intervals. When driving is adopted, recording is usually done by the operator.

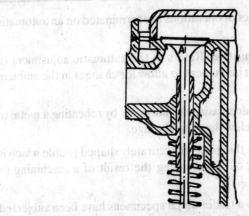

Tappet valve.

Tee-valve. A three-way screw-down stop valve having the connections in the form of the letter 'T'.

Teeth. Refers to the tooth-shaped projections on cog-wheels, gear-wheels, ratchet wheels, and many cutting tools.

Telechiric (telecheric) Device. A remotely controlled manipulator which is involving a human operator.

Telepresence. This term refers to robots which, at their worksite, have manipulations (end effectors) having sufficient dexterity to allow the operator to perform normal human functions; while at the control station, the operator receives an adequate sensory feedback to provide a feeling of actual presence at the worksite.

Teleprinter. A telegraph transmitter having a typewriter keyboard, by which characters of a message are transmitted electrically in combinations of five binary digits, being recorded similarly by the receiving instrument, which then translates the message mechanically into printed characters via control movements of an electric typewriter.

Telescopic Slide (telescopic shaft). Refers to two or more hollow tubes sliding one within another, providing a long support when extended.

Tell-tale. Refers to an indicator to show the amount of movement of a winding engine or of some mechanism, or to indicate the precise time when a series of operations has terminated on an automatic machine.

Temperature Compensation. Refers to the automatic adjustment of the reading of an instrument to allow for changes in the ambient temperature.

Tempering. Refers to decreasing the hardness by reheating a metal to a temperature below the critical range.

Template (templet). A thin plate of accurately shaped profile which is used for making out or checking the result of a machining or other operation.

Tensile Test. Refers to a test in which specimens have been subjected to an increasing load in tension, usually unit they break. A stress-strain curve plotted from the results has been used to determine the limit of proportionality; proof stress; yield point; ultimate tensile stress; elongation and area reduction of a specimen are also measured.

Tensile-testing Machine. A machine which is used for applying either a tensile or a compressive load to a test-piece, hydraulically or by power driven screws.

Tension Spring. A helical spring which is designed to extend with increasing load.

Tensioner. A spring which is loaded jockey pulley.

Terotechnology. Refers to the installation, commissioning, maintenance, replacement and removal of plant machinery and equipment feedback to design and iperation thereof together with the related subjects and practices.

Tesla (T). The unit of flux density = 1 Wb/m^2.

Test Load. See proof load.

Test Piece. An accurately made piece of material which is used for a tensile test, impact test or other testing machine.

Testing Machine. A machine which is used for applying loads to a test-piece to measure extension, etc.

Thermal Effciency. Refers to the ratio of the work done by a heat-engine to the mechanical equivalent of the heat supplied in the steam or fuel.

Thermal Stress. Refers to stress in a structure or mechanism which is caused by unequal expansion of different parts due to differential heating.

Third Tap. A plug tap.

Thompson Indicator. The Thompson indicator enlarges the movement BB' to AA' through simple levers and is used in engine indicators.

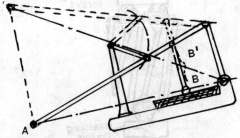

Thompson indicator-straight line motion.

Thoroughfare Tap. A tap having its square head small enough to allow it to pass through the hole which it has tapped.

Thrasher (thrasing machine, thresher). A machine which is able to separate the grain of wheat or other cereals from the straw and chaff. It has beating bars in a rotary cylinder or drum for separating the grain which is winnowed by means of shaking riddles or screens under the action of an air blast and finally delivered as clean wheat into a sack. The machine's name has been also spelt 'thresher'.

Thread Grinding. The accurate finishing of screw threads by a specially profiled grinding wheel which has been automatically traversed along the revolving work.

Thread Insert (aero-, wire-thread insert. Refers to a wire of approximately rhombic cross-section which is formed into a spring-like helix. It is inserted into tapped holes and retains another threaded member. Insertion has been accomplished with a driving tool engaging on the leading end tang. After full insertion the tang has been broken off at the notch leaving the insert just above the top of the countersink in the tapped hole. Thread inserts are used when a good resistance to wear is required for holes tapped in malleable materials, *e.g.*, threaded holes in light-alloy castings for aircraft where the weight of a more durable material has been prohibitive.

Thread Measurement. A simple method of measuring screw threads is shown in Figure. Three pieces of wire of known diameter have been inserted between the measuring contacts.

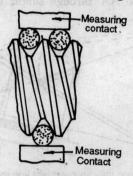

Three-wire system of thread measurement.

Thread Rolling. Refers to the formation of screw threads by rolling pressure.

Three-throw Crank. A crankshaft which is often used for pumps, with three cranks at 120° for driving three valves buckets or pistons.

Three-throw Pump (treble-barrel pump). A pump having three working barrels in line, having the piston rods connected to a three-throw crank.

Thresher. See thresher.

Threshold Stress. Refers to the valve of stress at the fatigue limit.

Throttle. Refers to the lever, or switch which is used for controlling the speed of an engine, by regulating the flow of the fuel-air mixture of the fuel or of steam in a steam-engine.

Throttle Governing. Governing a steam-engine by the throttle valve instead of by the shaft to the slide valve.

Throttle Valve

(*a*) Refers to the butterfly valve of a petrol engine.

(*b*) Refers to the regulator valve of a gas engine.

(c) Refers to the governor-controlled steam valve of steam-engines and turbines.

(d) The regulating valve controlling the pressure and temperature range of the working fluid in a refrigerator.

Throtting. Reducing the pressure of a fluid by causing it to pass through minute or tortuous passages.

Throw. Refers to the eccentricity of a crank or eccentric which is equal to twice the radius.

Throw Disc. Refers to the disc of a slotting machine which actuates the ram by a short connecting-rod.

Throw-out Gear. Refers to the reversing gear of a marine engine employing only a single eccentric.

Thrust

(a) Refers to the component of the resultant force from a propeller parallel to the propeller axis.

(b) Refers to the resultant force from a jet-engine or from a rocket-engine.

(c) Refers to the compressive force in a member of a structure, of an engine or for a mechanism.

Thrust Bearing (thrust block). Refers to a bearing on a shaft for taking an axial load (thrust) such as a ball-bearing with lateral races, a Michell bearing or a plain bearing pad.

Thrust Block. Thrust bearing.

Thrust Collars. Refers to collars on a shaft or spindle which transmit thrust to a thrust bearing.

Thrust Reverser. Thrust spoiler.

Thrust Shaft. Refers to a separate length of shafting on which have been formed the collars for the thrust bearing or a marine engine.

Thrust Spoiler (thrust reverser). A controllable device mounted at, or on, a propelling nozzle to reduce or to reverse the jet thrust.

Tie (tie rods), (tension rods)

(a) The rods which are supporting the jib of a crane. A tie.

(b) A purely tensile structural member.

Tilt Angle. Refers to the angle by which a carriage, using advanced passsenger train suspension gets titled relative to a line perpendicular to its wheel axis.

Time and Motion Study. Refers to an analysis of the timing and movements of an operator, or of parts of a machine (or mechanism) to ascertain how time can be saved or movements simplified when undertaking a particular operation on a piece of work.

Time Constant. Refers to the interval during which the value of the amplitude of a vibration, that decreases exponentially, decreases to one/e of its initial value.

Timing Chain. Refers to the chain between the sprocket wheels on the crankshaft and the camshaft of an internal-combustion engine.

Timing Gear. Refers to the two-to-one drive connecting the camshaft of a single camshaft engine to its crankshaft.

Timing Washers. Thin washers which are placed under the heads of the screws of a balance to alter its amount of inertia and thus its time of vibration.

Timing Wheels. Refers to toothed wheels on the crankshaft and a camshaft of a motor-vehicle which are connected by a timing chain to give a reduction ratio of one to two for a single camshaft engine.

Tip. Refers to the edge where the tooth flank meets the tooth crest.

Tip Angle. Refers to the angle between the axis of a bevel gear and the tip cone generator.

Tip-path Plane. Refers to the plane of rotation of the blade tips of a rotorcraft in flight; it is higher than the rotor hub.

Toe. Refers to the lower end of a vertical spindle working in a footstep.

Toe-in. Refers to the small forward convergence of the planes of the front wheels of a motor-vehicle to promote steering stability and to equalize tyre wear.

Toe-out. Refers to the outward inclination of the front wheels of an automobile on turns due to setting the steering-arms at an angle.

Toggle-joint (knee). A lever knuckle-joint which consists of two lever arms forming an angle with each other and hinged at the centre. The tree end of one lever has been hinged on a fixed pivot and that of the other lever has been tree to move. Any movement to bring the levers nearly into line causes a very large pressure to be exerted at the ends.

Toggle Press. A power press which is incorporating a toggle-joint.

Tolerance. An acceptable range of variation of some dimension.

Limits of tolerance (limits). Refers to the maximum by which the actual size of a dimension or of a profile, etc., are permitted to depart from the design size or form.

Bilateral tolerance. Refers to a tolerance which is allowing variation in both directions from the design size or form.

Feature tolerance. Refers to a tolerance on size of a particular feature such as a pin diameter or a slot width.

Form tolerance. Refers to the total variation allowed for the form of a feature, shape or profile.

Metal tolerance. Refers to the total dimensional variation, measured normal to the profile, in the amount of metal allowed on the surface of that profile, normally indicated by the maximum positive and/or negative departures from the design form.

Positional tolerance. Refers to the total amount of variation which is allowed for a positional feature, such as tolerances on distances between centres. These tolerances are normally distributed bilaterally or in all directions round centre.

Unilateral tolerance. A variation which is allowed in only one direction from the design size or form, such as in the dimensions of mating features.

Tongs Lifting

    (*a*) Shows lifting tongs carrying a plate in jaw.

    (*b*) Shows the jaws open, held up by *d*, and operated by a, through hinge *b* and chain *c*, and pivoted about *e*. This type of tong is used for rough bulky material.

Tonne (t). A unit $= 10^3$ kg.

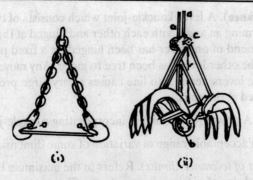

Lifting tongs: (i) carrying a plate (ii) tong jaws open.

Tool. Refers to any implement by which mechanical operations are performed, whether by hand or machine such as those used for forming and cutting.

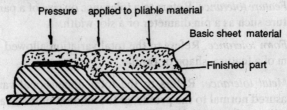

Pressure   applied to pliable material

Basic sheet material

Finished part

Combined blanking and forming tools.

Tool Carriage. Refers to the sliding carriage which carries the cutting tool in any automatically operated machine.

Tool-holder (cutter bar). The bar which is holding the cutting tools in metal turning and shaping.

Tool Post

    (*a*)   A clamp by which a tool has been held in the slide rest of a lathe.

    (*b*)   A clamp by which a tool has been held in the ram of a shaping machine.

Tool Room. Refers to that part of the machine shop having specialized machines, such as jig borers, where the manufacturing tools and jigs are made for the rest of the machine shop.

Tool Stay. A slotted bar which is held in the socket of the T- rest and with the slot embracing the flattered shank of the boring tool.

Tooling. Refers to the cutting of metals by cutters as opposed to the shaping of surface by grinding.

Tooth. See teeth.

Tooth Crest. Refers to the surface joining the two flanks of a single gear-tooth.

Tooth Flank. Refers to that portion of a tooth surface which lies within the working depth.

Tooth Profile. Refers to the line of intersection of the tooth flank with a defined surface.

Tooth Space

    (*a*) Refers to the difference between the radii of the tip cylinder and root cylinder of a cylinder gear.

    (*b*) Refers to the shortest distance between the tip cone and the root cone measured along the back cone generator of a bevel gear.

Tooth Trace. Refers to the line of intersection of the tooth flank with the reference cylinder of pitch cone.

Top Card. An indicator card which is taken from the top of a vertical or oscillating cylinder.

Top Steam. The steam which is entering above the piston in a double-acting steam hammer.

Torque. Refers to the turning moment about an axis, for example, of the air forces on a propeller, which may be uniform or fluctuating or of the turning moment of an engine crankshaft.

Torque Coefficient (of propeller). May be defined as the torque divided by the product of the density of the fluid, the square of the rotational speed and the fifth power of the diameter.

Torque Converter

    (*a*) Any device which acts as an infinitely variable gear, but with varying efficiency such as a fluid flywheel or coupling.

    (*b*) Any device which multiplies torque, for example, a gearbox.

Torque Dynamometer. A piece of test equipment which is used for absorbing the power and measuring the torque of an engine.

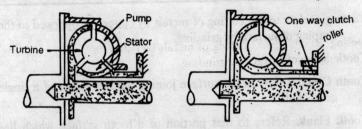

Torque converter   Torque converter coupling

Torque Limiter. A device which is used to prevent the torque of a constant-power turboprop from exceeding a certain value.

Torque Meter. A piece of test equipment attached to a rotating shaft which measures the angle of twist of a known length of shaft between two gauge points and thus enables the transmitted power to be calculated.

Torque Tube. Refers to a tube which is used to transmit or resist a torque.

Torsiograph. An instrument which is used for measuring and recording the amplitude and frequency of the torsional vibrations in a shaft.

Torsion. Refers to the application of a twist without any bending.

Torsion Bar Suspension. Refers to a springing system which is adopted in some motorvehicles, in which straight bars anchored at one end have been subjected to torsion by the weight of the vehichle.

Torsional Centre (centre of twist). If a couple has been applied at a given cross-section of a straight member, then the section rotates about a point in the plane of the section. This point, the torsional centre, remains stationary. It may be coincident with the flexural centre but has been not necessarily true.

Toughness. Refers to a condition which is intermediate between brittleness and softness, and indicated by a high ultimate tensile stress. With a low to moderate elongation and area reduction of a test-pie.e, plus a high value in a notched-bar test.

Track

   (a)   Refers to the rail or rails along which trains travel.

(b) Refers to the distance between the wheels of a vehicle, the wheels being on the same axle.

(c) Refers to the distance between the outer points of contact of the main wheels of an aircraft with the ground.

(d) Refers to the distance between the centre-lines of paired wheels in an aircraft undercarriage.

Track Rod. A rod which is connecting through ball joints the arms carried by the stub axles of a motor-vehicle so as to convey the angular motion from that wheel which is being directly steered to the other.

Traction. The propulsion of vehicles by power by virtue of the frictional adhesion of their wheels.

Traction Engine (road locomotive). A power-driven vehicle which is frequently propelled by steam, with large ribbed wheels for travel on roads, the wheels being gear-driven from a simple or compound engine mounted on top of the boiler. A rope drum has been provided for haulage purposes.

Tractor

(a) A vehicle which is used on a farm especially for drawing wagons and powering agricultural machines.

(b) An aeroplane with a power plant producing tension in the propeller shaft.

(c) Any mobile prime mover used for towing purposes on land.

Trail. Refers to the distance of the point of contact of a steered wheel with the ground behind that where the line of the swivel-pin axis intersects the ground.

Trailing Axle. Refers to the rearmost axle of a locomotive.

Trailing Edge. Refers to that edge of a wing, aerofoil, strut or propeller blade which last touches the air or water when the craft is in motion.

Trailing Lengths. Refers to the coupling rods of a fully coupled locomotive extending backwards to the trailing wheels.

Trailing Wheels. The wheels on the rearmost axle of a locomotive.

Train. Refers to a series of connected wheels or parts in machinery.

Compound train. A train of gearing in which intermediate shafts and gears are used obtain the desired torque or speed change within a small volume.

Train Brake. See vacuum brake.

Train of Gearing. Refers to a system of wheels and pinions providing an increase in torque with a reduction in speed or vice versa.

Tram. A wheeled vehicle running on rails.

Tram Wheel. A flanged wheel.

Transducer. A device which is used for converting a physical quantity (or signal) of one kind into another kind.

Transfer Line. A series of machines which are operating automatically on a continuous line of parts.

Transient. Refers to a non-steady motion impressed on a system which usually decays to a negligible value, but the lasting effect of the transient may be a different steady state.

Transmissibility. May be defined as the ratio of the amplitude of vibration of a specified particle within mechanical system to the amplitude of an applied vibration at a given frequency; normally the system is supported elastically, as by isolators.

Transmission. A term applied to various methods of transmitting and transforming power as by (*a*) line shaft, (*b*) belts and pulleys, and (*c*) gears.

Transmission Dynamometer. A device which is used to measure the power transmitted by a shaft either.

(*a*) By using a torque meter to measure the twist over a given length of shaft, or

(*b*) By direct measurement of the torque acting on an interposed differential gear.

Transverse Pitch. Refers to the distance between the traces of adjacent teeth of a gear measused round the reference circle.

Transverse Plane. Refers to a plane at an angle to the axis of a screw, gear, of body, usually, normal to the axis.

Transverse Pressure Angle. Refers to the acute angle between the normal to a tooth profile in a transverse section, where it intersects the reference circle and the tangent to the reference circle at that point.

Working transverse pressure angle. Refers to the angle with reference to the pitch circle instead of the reference circle.

Travelator. Refers to a moving platform which is constructed on the same principle as an escalator, for conveying passengers up or down a long slope, but without steps.

Traveller. A samll C-shaped spring clipped upon the ring of a ring spanner frame to act as a thread guide and to assist in the insertion of the yarn twist.

Travelling Bridge. A movable bridge which can be rolled backwards and forwards across an opening. Fig. shows a bridge carrying granular material which is shipped from a hopper D into trucks E. The carrier C picks material from the trough B into which it has been deposited from truck A.

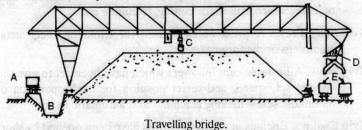

Travelling bridge.

Travelling Geer. Refers to the actuating gear of a travelling crane.

Traversing. A longitudinal motion of a cutter on a lathe or of any tool on a machine.

Treadle

    (*a*)   A foot lever which is connected by a rod to a crank to give motion to a lathe. Sewing machine or other mechanism.

    (*b*)   A contact which is operated by the deflection of running rails due to the passage of a train, trucks etc.

Treadmill. Refers to an appliance for producing motion by the stepping of a man or horse, on the steps of a revolving cylinder.

Treble Ported Slide Valve. An exhaust relief valve in a cylinder having two narrow ports in the body of the valve in addition to the end supply.

Treble Valve Box. A pump casing containing suction, delivery and intermediate check, or retaining valves.

T-rest. A T-shaped rest to support a tool which gets clamed to the bed of a wood-turning lathe.

Tribology. Refers to the study of friction, lubrication and wear.

Trick Valve (allan valve). A slide valve which is housing an internal steam passage additional to the exhaust cavity.

Trigger. A device, coarser than a flirt, for releasing a spring or catch and thus setting a mechanism in motion.

Trim. A term relating to the materials of the disc, body seat ring and stem of valves and stating the percentage of some element of the alloy of which the part is made, such as '10 per cent chrome trim'.

Trimming Machine (guillotine). A lever-or treadle-operated machine which is used for cutting, trimming etc.

Trimming Press. A press which is used for trimming sheet-metal stampings or die forgings.

Trip Dogs. Adjustable cam followers which have been set to give the correct set, speed, and turret position for each operation of automatic screw-making machines and automatic lathes.

Trip Engines. Engines in which the valves have been opened by short levers instead of eccentrics.

Trip Gear. A gear which is used for actuating valves, opening them by a trigger mechanism which, is then tripped out of engagement to allow the valve to close under a heavy spring.

Trip Lever. A bell-crank lever for the rapid opening and closing of valves.

Triple-expansion (steam) Engine. Refers to a steam-engine with high pressure, intermediate pressure and low-pressure cylinders working on the same crankshaft.

Tripping. Refers to the running of a tooth of an escape wheel past the locking face in an escapement.

Trolley (trolly)

   (*a*) A low truck which is running on rails.

(*b*) A small table which is running on wheels.

(*c*) A truck that can be titled.

Trolley-bus. An electrically-driven omnibus which is collecting current from an overhead system of wires by means of a pole and trolley wheel.

Trolley Wheel

(*a*) A single or double-flanged wheel on a trolley.

(*b*) A small grooved wheel for collecting current from an overhead wire for a trolley-bus.

Trunnion (trunnion bearing). A bearing on which a vessel or cylinder can rotate or oscillate.

Trussed Shaft. A long shaft which has been supported by rods to provide rigidity.

Tube

(*a*) A rigid or flexible pipe which is used for transmitting fluids or slurries usually in some permanent application.

(*b*) A structural member of hollow, usually circular, section; a hollow strut.

Tube Mill. A ball mill having a cylinder longer than usual; this usually being subdivide internally so that the material to be ground passes from one compartment to next, the grinding media in successive departments being progressively and appropriately smaller.

Tumbler Bearing. A support bearing for long shafts which, when the carriage, or traveller, comes into contact with it, pivots on an external support and allows the carriage to pass, but returns immediately to its poisition when released.

Tumbler Gear. Refers to a system of four gears arranged so that the direction of the driven shaft may get reversed by the inter position of an extra wheel in the gear train. A common application, for reversing the direction of the lead screw in a lathe, has been achieved by mounting two meshed idler gears on an 'L' shaped support pivoting about the axis of the driven gear, with which they are also in mesh: A small rotation of the support brings either one or the other of the idler gears into mesh with the driving gear forming a three or four gear train.

Tumbler Lock. A lock in which a tumbler or latch engages with notches in the bolt of the lock thus preventing its motion until the tumbler has been lifted or displaced by the key removing the obstacle and then shifting the bolt.

Tumbling. Refers to the process of revolving workpieces in a barrel with abrasive and other material for the purpose of deburring, cleaning and improving the surface finish or changing the lustre.

Turbine. It is a rotary power unit which is driven by the impact of, or reaction from, a flowing steam of air (air turbine) of hot gases (gas turbine), of water (water turbine), or of steam (steam turbine) on the blades, buckets or vanes.

Turbine Disc (or drum). Refers to rotating member on which the blades of the turbine are fixed.

Turbine Pump. A rotary pump with a number of stages which can thus lift higher heads than a centrifugal pump.

Turbine Rotor. See rotor.

Turbine-type Axial Compressor. Fig. shows the stator and rotor of a turbine-type axial compressor.

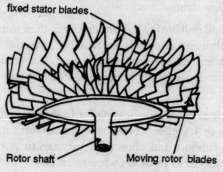

Turbine-type axial compressor.

Turbine Wheel, Turbine Rotor. The fundamental Hero type of turbine wheel as used in lawn sprinklers is shown Figure.

Turbo-generator. Refers to a directly coupled steam-turbine and generator.

Turbojet Engine. Refers to an air-swallowing engine, composed of a compressor, combustion chambers and a gas turbine, which

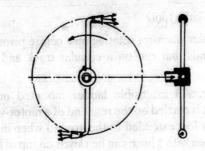

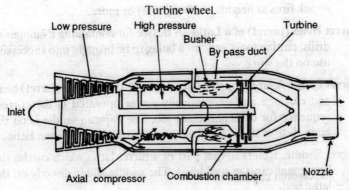

Turbojet engine (by-pass type).

generates thrust by a jet of hot gases passing down an exhaust cone to a propulsion nozzle. Shows the by-pass type.

Turboprop Engine. Refers to an air-swallowing turbojet engine which is coupled to a propeller and providing thrust mainly by the propeller and partly by a jet.

Turbopump

(a) A centrifugal pump with guide vanes at the exit from the impeller.

(b) A combined ram-air turbine and hydraulic pump for a guided weapon.

Turbo Starter. An independent turbine which is driven by compressed air, a gas source or other means and used for starting an aircraft engine.

Turning Circle. The circle of minimum radius in which a vehicle can be turned.

Turning Tool. See lathe tool.

Turntable. A circular platform rotating on a centre pivot with wheels under the ends that run on a circular track and with rails a diameter.

Turntable Ladder. An extensible ladder mounted on a rotating platform that is carried on the rear end of a motor-vehicle and is supported by feet extended to the ground when in use. By this means, firemen with a hose can be raised on top of the ladder to attack fires at heights of 20 m (60 ft) or more.

Turret Head (turret) of a Lathe. A device for containing a number of drills, cutting tools, etc., on a lathe, to be brought into successive use on the work.

Turret Lathe. A large capstan lathe in which the capstan (turret) head and carriage are automatically power-operated in the correct sequence for each particular job. The capstan saddle is on the main slide of the bed which is not the case on a capstan lathe.

Turret Saddle. Refers to that part of a turret lathe which carries the hexagonal capstan tool head. The saddle slides directly on the lathe bed.

Twin Screws. A pair of right and left-handed propellers on separate parallel shafts.

Twin-shaft Turbine. Two similar turbines mounted on the same horizontal shaft but discharging in opposite directions and hence balancing the end thrust.

Twist Drill. A hardened steel drill with cutting edges formed at the periphery of helical flutes and with a conical point backed off to provide clearance for the borings as they are cut of.

Twist Drill Grinder. A grinding machine set for grinding the constant angle of a twist drill.

Two-high Rolls (two-high mill). A rolling mill with two rolls only, one above the other.

Two-stroke (2 stroke) Cycle. A piston-engine cycle completed in two strokes, involving one crankshaft revolution, the charge being introduced, compressed, expanded and exhausted through ports in the wall of the cylinder, before and during the entry of the fresh charge.

Two-throw Crank. Refers to an axle of shaft with two cranks, usually at right-angles to each other.

Tyre. Refers to a forged and flanged steel ring shrunk on the rim of a locomotive or other wheel to give added strength and durability.

Tyre Rolling Mills. Vertical or horizontal mills in which the tyres have been expanded and shaped been inner and other rolls in a roughing and a finishing pass.

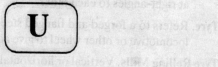

U. A symbol for strain energy.

Ultimate Load

    (*a*)   Refers to the maximum load which a structure is designed to withstand without a failure.

    (*b*)   Refers to the product of the limit load and the ultimate factor of safety.

Ultimate (tensile) Strength. Refers to the breaking load under tension.

Ultimate Tensile Stress. Refers to the ratio of the highest load applied to a piece of metal during a tensile test divided by the original cross-sectional area. Also known as 'tenacity'.

Ultrasonic Testing. Refers to the use of very high frequency sound waves to investigate the continuity of the material of a work piece.

Unaflow Engine (uniflow engine). Refers to a steam engine in which the steam enters through drop valves at the ends of the cylinder and exhausts through a piston-controlled belt of ports at the centre.

Under Frame. Refers to that part of a truck having the axles and their bearing springs, the axle boxes, the buffer and drawbar springs, and the wheels.

Under Carriage (landing gear). Refers to the assembly of wheels brakes and shock-absorbing device fitted beneath an aircraft which allow it to land and support the aircraft when on the ground. Most aircraft retract their undercarriages during flight.

Undulation. Means a periodic departure, or departures, of the actual tooth surface of a gear from the design surface due to machining or other variations.

    Undulation height. Means the normal distance between the crests and troughs of tooth undulations.

Undulation wavelength. Means the distance between two adjacent crests of an undulation.

Ungeared (machine)

(a) A lathe or drilling machine without a back gear.

(b) A direct drive as from an aircraft engine to a propeller without gearing.

Unit Stress. Refers to the stress upon a given sectional area per square inch, square foot or square metre ($1\,Pa1 = N/m^2$).

Universal Chuck (concentric chuck). It is same as self-centring chuck.

Universal Joint. The term for a joint between two shafts which have been not necessarily in line and whose axial position has been not necessarily stationary. The joint may partly be made of rubber between yokes on the adjacent shafts or it may consists of purpose-made metallic parts. A Hooke's joint is kept at both ends of the propeller shaft in a automobile to allow for movement of the driven axle relative to the gearbox.

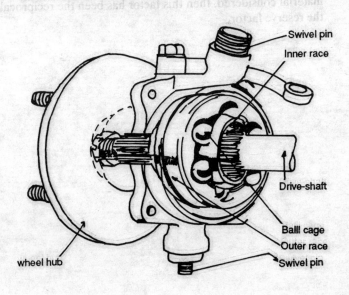

Universal joint.

Universal Transfer Device. An industrial robot.

Universal Vice. A vice which can be turned on different axes and fixed in almost any desired position. Its movable jaws can be positioned by a hand screw or pneumatically for quick action.

Unstable Equilibrium. See equalibrium.

Up-and-down Indicator. Refers to mechanism for indicating when a chronometer or watch needs, 'Up' indicating when fully wound.

Upset Forging. Refers to a process in which the steel is kept against a stationary die impression, a movable grip die moves in to hold the piece and a plunger advances to form the head on a bar or other piece. Three times the diameter has been the maximum obtainable without buckling. The process was originally designed for producing bolt heads but has been developed for other operations.

Utilization Factor. Refers to the ratio of the in-service stress to ultimate stress. If stress has been proportional to load for the material considered, then this factor has been the reciprocal of the reserve **factor**.

V. The symbol of potential energy.

VHN. Vickers Hardness Number.

Vacuum Brake. Refers to a brake system on passenger trains. A vacuum maintained in reservoirs by exhaust pumps, has been simultaneously applied to brake cylinders throughout the train. When the vacuum is broken all brakes are automatically applied.

Vacuum Gauge. A gauge which indicates the amount of vacuum in a vessel partially evacuated of its air or in a steam condenser.

Valve. A lid or cover to an aperture that opens a communication for a liquid or gas in one direction or closes it in another, or regulates the amount of flow, either manually or automatically.

Valvebox (valve chamber, valve chest)

    (*a*) The chamber having the valves or valve in a fore pump or steam-engine.

    (*b*) The steam chest having the slide-valve.

Valve Chamber. Valve-box.

Valve Cock

    (*a*) A cock which opens with a lift valve or by means of a slide valve.

Valve Diagram. A graphical method for correlating the movements of the eccentric, of the valve and of the points of admission, cut-off, compression and release for the slide valve of steam-engine.

Valve, Four-way Piston. Figure shows the working of a four-way piston valve with by-pass control.

Valve Gear. Refers to the mechanical arrangements for actuating valves.

Valve-gear (overhead mushroom). Fig. shows the mechanism of an overhead mushroom valve-gear as fitted on motor cycles.

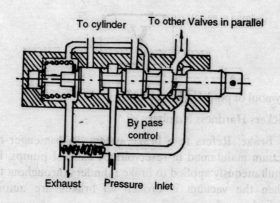

To cylinder   To other Valves in parallel

By pass control

Exhaust   Pressure   Inlet

Four-way piston valve.

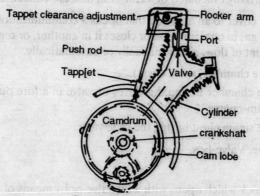

Tappet clearance adjustment   Rocker arm

Port

Push rod

Tappet   Valve

Cylinder

Camdrum

crankshaft

Cam lobe

Overhead mushroom valve-gear.

Valve Insert. A valve seating of special steel which is pressed into the heads of high-duty petrol engines.

Valve Lap. The distance moved by the piston from the central position to the point where the port pressure has been a maximum.

Valve Milling Machine. A milling machine for milling square or hexagonal portions of valves and cocks.

Valve-opening Diagram. A diagram which shows the opening area of a valve plotted against the internal-combustion engine crank angle or piston displacement.

Valve Port. Refers to a port in the valve body of a piston valve.

Valve Rocker (rocker arm). A small pivoted lever which transmits motion from a cam or a push rod to a valve stem.

Valve Rod (valve spindle). The rod to which a valve gets attached.

Valve Sector

(a) A slot link.

(b) The vertical sliding link of an oscillating cylinder engine which communicates the eccentric's motion to the valve rod weigh shaft.

Valve Spindle. Valve rod.

Valve Spring

(a) A helical spring which is used to close a poppet valve after it has been lifted.

(b) Refers to spring for closing a valve that has been lifted mechanically or by fluid pressrue.

(c) A spring which is used to force the packing rings of a slide valve against its working face.

Valve Yoke. The briddle of a valve rod.

Vane (air vane). A pivoted free surface which turns along the wind direction or a fixed curved surface which changes the direction of flow.

Vane

(a) Refers to a pivoted free surface which turns along the wind direction like a weather cock.

(b) Refers to a fixed curved surface for changing the direction of flow of a liquid or gas.

(c) The bucket of a turbine.

Vanish Cone. Refers to the cone whose surface passes through the roots of the washout thread.

Vapour Blast Cleaning. Blasting the surface of a workpiece with a high-velocity jet of vapour to give a smooth clean surface.

Vapour Lock. Refers to the formation of a vapour in a fuel pipe, resulting in the interruption of flow.

Variable-area Propelling Nozzle. A turbojet propelling nozzle with variable outlet area to match the different operating conditions especially with an after-burner. It may be actuated mechanically or aerodynamically, called 'eyelid' or 'petal' type, these names being self-descriptive.

Variable Cut-off. Refers to a cut-off actuated from a governor which is brought into action according to the load on the engine.

Variable Expansion. Refers to the expansion of steam in a steam-engine where the amount alters under the varing conditions of working, automatic or otherwise.

Variable Gears. Toothed wheels which are able to transmit varying velocity ratios during the course of a single revolution.

Variable-pitch Propeller. A propeller, the pitch of whose blades can be varied in fight.

Variable-speed Gear. A device, consisting of smooth speed cones or expansion, pulleys, whereby the speed ratio between two shafts can be varied without shifting the belt.

Variable Stroke (variable capacity engine). See engine (servo- motor types).

V-belts. Belts with a cross-section of vee shape for use with expansion pulleys, etc. A typical cross-section is shown in figure.

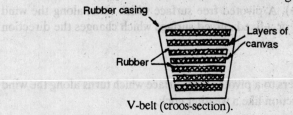

V-belt (croos-section).

Vector. A vector is a directed line segment that having both magnitude and direction. For example, velocity, a vector, describes how fast and in which direction; while speed, a scalar does not describe in which direction the object is moving, velocity, force and impulse are all examples of vector quantities.

Vee-thread (angular thread). A screw thread having a vee-shaped profile.

Velocity. The rate of change of position, or rate of displacement of a point with respect to a specified reference frame of co-ordinates and expressed in feet (or centimetres) per second. The direction as well as the magnitude are needed for the complete specification of a velocity.

Velocity Ratio

(a) Refers to the ratio of the distance moved through by the point of application to the corresponding distance moved by the load.

(b) Refers to the ratio of the velocities of bodies which are mutually connected by gearing, etc.

Velox Boiler. A forced circulation boiler giving very high rates of heating, having a supercharged furnace and a high gas velocity, which produce a high effciency for low weight.

Vernier. A movable auxiliary scale, sliding in contact with the main scale of graduation, to enable readings to be made to a fraction (*e.g.*, a tenth) of a division in the main scale.

A vernier calliper for measuring internal and external dimensions is illustrated in (a).

Vertical Boring Mill. A mill in which the work is held on a revolving horizontal table and the tools held in a turret on a single up-right head or on a cross-rail spanning two uprights.

Vertical Engine. Refers to an engine with the cylinders arranged vertically above the crankshaft.

Vertical Milling Machine. A milling machine with a vertical spindle, as found in end mills and often used for profiling.

Vertical Shaping Machine. Slotting machine.

Vibrating Conveyor. A trough which gets attached to vibrators which impart an upward and reciprocating movement to propel material upward and forward during the upward movement and which draw back the trough underneath the material in the subsequent movement.

Vibrating Links. Suspension links.

Vibration. A repetition of some value of velocity in an arbitarily selected particle in a mechanism.

Vibration Isolator. A resilient mounting which is used to absorb the vibration of a fixed mechanical installation.

Vibration-testing Machine. A machine which is used for subjecting specimens to desired frequencies and amplitudes of vibration to determine their operational characteristics. These machines have been of three main types:

(*a*) Wholly mechanical using cams or cranks.

(*b*) Electromagnetic using a moving-coil transducer to give linear vibrations.

(*c*) Hydraulic linear actuator incorporating a servo valve of which a number can easily be linked together.

Vibratory Feeder. A feeder delivering material in accurate quantities down an inclined surface, controlled magnetically or by vibrating the surface.

Vibrometer. An instrument which is used for indicating variations from the correct balancing of revolving machinery.

Vice (vise). A workshop tool for holding or gripping work firmly, consisting of a pair of steel-faced jaws, one of which is movable by a screw or by a lever; the other jaw is fixed.

Vickers Hardness Number (VHN). A number equal to 0.927 p, where p denotes the yield pressure, the area on which it acts being 0.927 times that of the surface area of the contacting faces. The pyramidal indenter is square-based and the opposite faces contain an angle of 135°. The Vickers and Brinell hardness numbers for a given load are nearly equal.

Virtual Work. Means an increase in potential energy due to the work done on an item by the external forces applied to it.

Viscosity. May be defined às the internal friction due to molecular cohesion in a field; the resistance to the sliding motion of adjacent layers of a fluid when in motion.

The 'coefficient of viscosity' may be defined as the tangential force per unit area necessary to maintain unit relative velocity between two parallel planes which are unit distance apart. It is inversely proportional to temperature. Units; poise.

The 'kinematic viscosity may be defined as the viscosity of the fluid divided by its density.

W

ω. Greek Letter omega. The symbol for angular velocity.

Wabblers. The coupling boxes which are connecting the breaking pieces with the necks of the puddling rolls.

Wallow Wheel. A bevel wheel on a vertical shaft with the teeth facing downwards.

Walschaert's Valve Gear (waldegg valve gear). Refers to a radial type gear driving the valve of a steam locomotive through a combination lever whose oscillation is the resultant of sine and cosine components of the piston motion.

Wankel Engine. Refers to an internal-combustion engine with a rotary piston, which is a three-lobe rotor that rotates in the combustion chamber to provide the same cycle as in a reciprocating piston engine.

Washer. Refers to an annular piece of metal, leather, rubber, etc., for distributing pressure under a nut or making a tight joint between surfaces shows different types of washers. (*a*) plain washer, (*b*) externally serrated, and (*c*) internally serrated lock washers, (*d*) taper washer for use on angle iron (e) helical spring lockwasher (f) tab washer (g) tab washer locking nut from turning (h) two-coil spring lockwasher.

Washing Down. Refers to thinning down to a feather edge.

Water Cylinder. Refers to the pump barrel of a steam pump.

Water Gauge
- (*a*) Refers to a vertical or inclined glass tube which is used to indicate the water level in a boiler.
- (*b*) A pressure described as the height of water supported by the pressure.

Water Jacketing. The casing of engine cylinders with a jacket through which water flows to keep them cool.

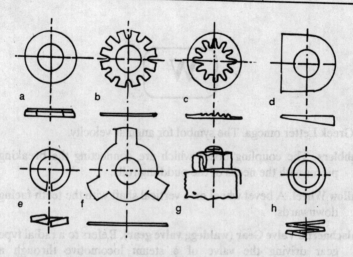

Washers.

Water-cooled Engine. An engine which is cooled by water jackets, typically water circulating within the cylinder block of an internal combustion engine.

Water-pressure Engines. Turbines, hydraulic rams, etc., which are driven by water pressure or by a head of water.

Watt (W). The unit of power = 1 J/s.

Watt Governor; (A pendulum governor). A pair of links, pivoted to a vertical spindle, term inate in heavy balls; shorter links have been pivoted to themid-points of these links and to the sleeve operating the engine throttle.

Watt's Straight-line Motion. A type of motion which is used by Watt to guide the piston rod in many of his early steam-engines. The point C in Fig. oscillates along a straight line as levers A and B rotate about their fixed pivots.

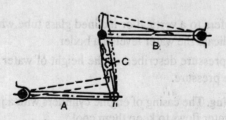

Watt's straight-line motion.

Waviness. Surface irregularities that have greater spacing than roughness.

The height is peak-to-valley distance and the width is the spacing of adjacent waves.

Weber (Wb). Refers to the unit of magnetic flux; it is the flux which, linking a coil of one turn, produces in it an e.m.f., of one volt as it is reduced to zero at a uniform rate in one second.

Wedge Gate Valve. A gate valve in which closure has been effected by the wedge action between the gate and the body seats, the gate being either solid or cored in one piece or in two pieces.

Wedge facing rings. Rings of different material from the wedges and secured to them, on which the wedge faces have been machined.

Wedge Gearing. Gearing which is composed of wheels with circumferential grooves which fit into each other and thus provide a friction drive.

Weigh Shaft (way shaft, reversing shaft). Refers to the shaft for moving the slot links to put a steam-engine into forward or backward gear.

Weighbridge. A table which is carried on a system of levers with arms so proportioned that a large weight on the weighbridge table can get balanced by a small weight moved along a steelyard.

Weighing Machine, Self-indicating. A weighing machine with a counterpoise in which the load is indicated by a movable pointer on a graduated scale as shown in Fig. The graduation is based on

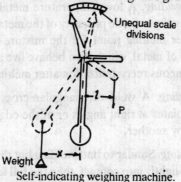

Self-indicating weighing machine.

the equality P. I = Wx, where P denotes the weight to be measured, and W denotes the counterpoise weight. In the simple form shown in the diagram the scale is not uniform; by using a flexible strip, moving on a cam, for P, the scale can be made uniform.

Weight. May be defined as the resultant force of attraction on the mass of a body when subjected to the gravitational field of another body. On Earth the units of weight have been based on a gravitational acceleration of 9.81 m/s^2 (32.174 ft/sec^2). A mass of 1 kg exerts a force (weight) of 1 newton when subjected to an acceleration of 1 m/sec^2.

Welding. Refers to the joining or fusion of pieces of metal by raising the temperature at the joint to make metal plastic so that the pieces can be joined or fused together.

Arc welding. Employs the heat of an electric arc to bring the metal to a molten state for joining by fusion.

Braze welding. Utilizes a filler rod that melts at a temperature greater than 700 K (800°F) but lower than the melting point of the base metal; it can be applied to all metals that melt above 800 K (1000°F), except aluminium and magnesium.

Butt weld. Refers to an end-to-end weld of two plates to form a continuous plate.

Electron beam welding. Using an electron beam for fusing the metal to make a joint.

Eutectic welding. A low-temperature metal welding process which is using the eutectic property of the metals involved, there being a lower melting point for the mixture of the particular combination of metal, which then behave live a pure compound with simultaneous recrystallization after melting.

Fillet welding. A weld of triangular cross-section which is joining two plates at right angles or at the edge of a thick plate overlapped by another.

Flash welding. Similar to butt welding but the parts have been first brought lightly into contact, then the full forging pressure is applied as electric are flashing commences.

Forge welding. Refers to the joining of two or more steel or iron parts by heating until they reach a plastic state forging them together.

Friction welding. A weld may be obtained by the heat generated by mechanical friction. For example, dissimilar metal shafts can be joined if one has been spun up against the other, with both being allowed to spin as the weld temperature is reached, also some bolts can be attached to plates by firing them with a special gun, with the impact temperature welding them on.

Gas welding. It uses an oxy-acetylene or oxy-hydrogen flame to get the desired temperature.

Heli-arc welding. Welding using helium as the inert gas.

Laser welding. Welding using a laser as the heat sources.

Plug weld. A series of holes drilled in one plate are fillet welded to join it to another over lapping undrilled plate. Alternatively, pins, jig pins or plugs may project from one part through holes in another to which they are then welded.

Projection welding. Points on one of the two mating surfaces to be welded have been left proud. The two parts are brought together under pressure and an electric current develops the necessary heat to weld all projecting points at one time.

Resistance seam weld. The overlapping plates have been joined by fusion as in a spot weld with the electric current flowing between two rotating wheels.

Resistance welding. Joins metals by the simultaneous application of pressure and heat.

Spot welding. It is the joining of two or more thin metal plates at a number of sports by local heating at these places with a heavy electric current for a short time.

Tig welding. Welding using a tungsten filler rod while the weld has been surrounded by a continuous flow of an inert gas.

Weldment. A welded assembly.

Westinghouse Brake. An air brake on railway rolling stock which is controlled by a reservoir of compressed air underneaths the engine and connected to a cylinder and piston under each carriage.

Wheel

(a) A solid disc or a structure which is made up of a rim and hub united sometimes by spokes, and capable of rotation on its own axis, such as a car wheel or pulley.

(b) Refers to the larger of an equal pair of gears. Equal gears have been usually called wheels if the diameter is greater than the width of the rim.

Wheel-base. Refers to the distance between the leading and trailing axles of a vehicle.

Wheel-and-disc Drive (gear). A ball-and-disc gear in which the roller and balls have been replaced by a wheel.

Wheel Dresser. Alternate plain and star-shaped hardened-steel discs which are mounted and freely revolving on the end of a long handle. The disc have been applied under pressure to the abrasive wheel that has to be dressed.

Wheel Plate. Quadrant plate.

Wheel-quartering Machine. A horizontal drilling machine with opposed spindles at opposite ends of the bed for drilling simultaneously the crank-pin holes in both wheels on a locomotive coupled-axle to ensure the precise angular relationship.

Wheel Track. Refers to the distance between two co-axial wheels often mounted on the same axle.

Wheeling Machine. A machine having one flat and one convex wheel for producing curved panels of sheet metals, or for finishing off a workpiece after panel beating by hand or machine.

Whip

(a) A whip crane.

(b) An arm of a windmill which is carrying the cross pieces and sails.

Whip Crane. See crane.

Whipping Drum. Refers to the winding barrel of a whip crane.

Whirling Arm. An apparatus which consists of a long horizontally rotating arm for carrying models for aerodynamic tests or for subjecting apparatus and men to high accelerations.

Whirling of Shaft. Refers to the deflection from straightness of a shaft at certain critical rotational (whirling) speeds.

Whirlpool Chamber (vortex chamber). Refers to the space surrounding the impeller of a centrifugal pump into which it discharges.

White Metal. An easily fusible alloy with a tin base (over 50%) having lead, antimony and copper, used for lining bearings.

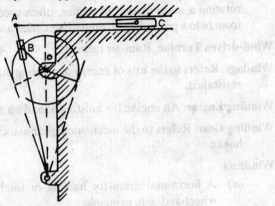

Whitworth quick-return motion (1).

Whitworth Quick-return Motion. Fig. shows a simple quick-return motion, the point B rotating about O causes the slide C connected through A to move slowly from left to right and return quickly as indicated by the relative lengths of the arcs of the circle.

Whitworth Screw Thread. Refers to a symmetrical vee-thread of 55° included angle; one-sixth of the sharp vee is truncated at top and bottom, the thread being rounded equally at crests and roots by circular arcs blending tangentially with the flanks, the theoretical depth of thread being thus 0.640327 times the nominal pitch.

Wicket Gate. Refers to a regular of the supply of water to water turbines having guide vanes pivoted at their centres which are automatically adjustable to the rotation speed of the turbine wheel and vane angles regulated by a pendulum governor acting through a servo-motor.

Winch

    (*a*)   Refers to a steam-engine on a ship's deck used for hoisting cargo, etc., with various drums and barrels driven by gearing at various speeds.

    (*b*)   Refers to a hoisting machine which is operated by hand or driven by power.

Wind Pump. A pump which is operated by the force of the wind rotating a multi-bladed propeller, often used for raising water from below ground level. Sometime called a windmill.

Wind-driven Turbine. Ram-air turbine.

Windage. Refers to the loss of energy of rotating machinery due to air resistance.

Winding Engine. An engine for hoisting a load up a shaft.

Winding Gear. Refers to the mechanical gear associated with lifts and hoists.

Windlass

    (*a*)   A horizontal drum for hauling or for hoisting, using the wheel-and-axle principle.

    (*b*)   An apparatus for hoisting an anchor by means of its cable and for lowering it with a brake to regualte the paying out of the cable.

Windmill. A mill worked by the action of wind on large sails mounted on whips, the whole apparatus being rotatable on top of the building and the angle of the sails adjustable.

Windmilling

    (*a*)   An aircraft's propeller which is rotating but delivering no power to the propeller shaft.

    (*b*)   A compressor which is rotating by the pressure of the airstream.

    (*c*)   The rotor of an autogiro (because it is not power driven) is windmilling; likewise, the rotor of a helicopter in case of engine failure.

Wing Valve. A conical-seated valve which is guided by radial vanes of ribs when fitting inside a circular port.

Wiper

    (a)   Refers to an oscillating bar which cleans the face of a car's windscreen.

    (b)   Refers to the cam teeth on the wheel of a tilt hammer.

    (c)   Refers to a mechanism used in weaving to convert a rotating into a reciprocating motion.

Wobble Crank. Refers to a short-throw crank for giving an elliptical motion to a sleeve valve by a short connecting-rod and ball-joint.

Wobler Test. Refers to a fatigue test of specimens of materials held with one end in a rotating chuck and carrying a weight on a ball-bearing at the other end.

Work

    (a)   Refers to the overcoming of resistance (force) through a certain distance. The units of work are the erg and the foot pound.

    (b)   a workpiece.

Work Hardening. Refers to an increase in strength and hardness of metals which is produced by making them do work, such as resisting stretching or forming. It is most pronounced when cold-working metals such as iron, copper, aluminium and nickel which do not recrystallize at room temperature.

Working Barrel. A pump barrel having the piston and bored portion and sometimes the clack valves.

Working Cylinder. The exploding cylinder in a gas-engine which is having a separate compressing cylinder.

Working Depth. Refers to the depth on the tooth face which the tooth of the mating gear extends, being less than the total length of tooth from point to root by the amount of the bottom clearance.

Working Load. Refers to the mean ordinary load to which a structure or mechanism is subjected.

Working Pressure. Refers to the pressure at which an apparatus or engine works, as distinct from its test pressure.

Working Space. The volumetric geometry of the space within which a particular robot has been capable of performing its work.

Working Stress. Refers to the safe stress for a structure or mechanism, based on experience and distinct from any proof stress.

Workshop Gauge (manufacturing gauge). A gauge which is used for checking during the actual production of the work on a machine.

Workshop Microscope. A machine mounted optical microscope which is used for observing threads, cutting tools and grinding wheels in situ.

Worm (worm gear, worm wheel). A helical gear that meshes with a worm wheel in sliding contact.

Worm Gearing. Gearing which is composed of worms and wheels. The worms normally drive the wheels and provide gears of high reduction ratio connecting shafts with axes at right angles, but whose axes do not interest.

Worm Thread. Refers to that portion of a worm bounced by the root cylinder, tip cylinder and the two helicoid surfaces.

Worm (screw) Conveyor. Refers to a conveyor in which a revolving worm continuously propels loose material along or through a tube.

Worm Wheel. A concave face gear-wheel having teeth capable of line contact with a worm.

Worm Wheel Hobbing Machine. Refers to a machine for cutting teeth using a hob which is the exact counterpart of the worm.

Worm-and-wheel Steering Gear. Refers to a gear in which the worm on a steering column meshes with a worm wheel or sector, the latter being attached to the spindle of a drop arm.

Wrap-up. A means for providing relative displacement between a driving and a driven member when a threshold value is exceeded, subsequently annulling it and restoring a positive drive when below that value, such as in the case of a given torque or of a given number of turns, etc.

Wringing. Refers to the process of sliding one slip gauge on another by a wiping motion to remove all air and dirt between the mating surfaces. When properly wrung together, the gauges cling to each other with negligible error.

Wrist. The extremity of a robot arm to which an end effector gets attached.

Wrist Articulation. Refers to the rotational degrees of freedom of a robot's wrist.

Wrist Plate (motion or rocking disc). The plate, attached to the side of the cylinder of a Corliss engine and worked by levers, which transmits motion by connecting-rods to the valve spindle.

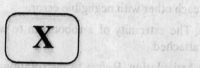

X-spring. Two superimposed laminated springs which are forming a letter X. It is used on carriages and some balances.

integral with or separate from the bonnet.

cross-sectional area to the

Y

YP. Yield Point.

Y-lever. Refers to the longest lever of a weighbridge to which the
steel-pard's rod gets attached.

Yale Lock. Refers to a cylinder lock for doors in which a key is raising
a number of springs to different heights for its release.

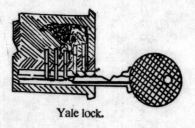

Yale lock.

Yankee Machine. Single-cylinder machine.

Yard Traveller. Outdoor overhead traveller.

Yaw. Refers to an angular movement of a robot's wrist about an axis
perpendicular to both the axis of the robot's arm and the
horizontal axis of the wrist.

Yield Point. May be defined is that point in the loading test on a
test-piece when the deformation increases suddenly and a
substantial amount of plastic deformation takes place under
constant (or reduced) load, for example in the stretching of a
test-piece in tension.

Yielding Attachment. Refers to a special method by which the outer
end of a mainspring gets attached to the barrel to all and
concentric uncoiling of the spring.

Yoke

    (a) The frame which is formed in a slide-valve spindle to
embrace the box-like portion of the vale and thus form the
connection.

(b) The 'Y'-shaped frame which connects the root-end of a solar panels to a spacecraft. When unfolded, the extended panels gets titled by a drive mechanism which rotates the tail of the 'Y' to keep the panels pointing at the Sun.

Yoke (of a valve). Refers to the exterior part of an outside screw valve in which the actuating thread of the stem engages either directly or through a bush or through a yoke sleeve. The yoke may be integral with or separate from the bonnet.

Young's Modulus (modulus of elasticity). May be defined as the ratio of the stretching force, as on a test specimen, per unit cross-sectional area to the elongation per unit length. Its value has been of the order of 10^5 Pascals or 10^7 Ibf/in^2 for metals.

Z

Zero (setting). Means the setting of the zero of an instrument before measurements are started.

Zerol Bevel Gear. A spiral bevel having curved teeth and having a zero degree mean spiral angle.